HOW TO RAISE
POULTRY

EVERYTHING YOU NEED TO KNOW

BREED GUIDE & SELECTION
PROPER CARE & HEALTHY FEEDING
BUILDING FACILITIES

Christine Heinrichs

Voyageur Press

DEDICATION

To all the poultry breeders who have helped me understand
their birds. They are truly a rare and precious breed.

Voyageur Press titles are also available at discounts in
bulk quantity for industrial or sales-promotional use. For
details write to Special Sales Manager at MBI Publishing
Company, 400 First Avenue North, Suite 300, Minneapolis,
MN 55401 USA.

To find out more about our books, visit us online at
www.voyageurpress.com.

Library of Congress Cataloging-in-Publication Data

Heinrichs, Christine.
  How to raise poultry / Christine Heinrichs. — 1st ed.
      p. cm.
  Includes bibliographical references and index.
  ISBN 978-0-7603-3479-9 (sb)
  1. Poultry. I. Title.
  SF487.H444 2009
  636.5—dc22

                                    2008031989

Edited by Danielle Ibister
Designed by Sara Holle

Printed in Singapore

CONTENTS

ACKNOWLEDGMENTS

Writing *How to Raise Chickens* several years ago was such a blessing. I felt a sense of unreality when Voyageur Press asked me to write *How to Raise Poultry*. With its broader range, it's a different kind of journey.

Inevitably, including a wide range of species rather than focusing on a single species results in more of a survey and summary than a detailed account. My hope is that readers will use this book as a guide to making decisions about which birds are right for them and use it to direct them as they learn more about birds that have not yet joined their flocks.

The book began during the National Tropical Botanical Garden (NTBG) Environmental Journalism Fellowship on Kauai, for which I was selected in 2007. The focus was on invasive species, a huge problem for the islands. Indigenous flora and fauna have few defenses against species brought in with the tourists who are the mainstay of the Hawaiian economy. It's been a long decline for native plants and animals since Captain Cook arrived in 1778.

As we studied plants at the garden, we inevitably discussed animals. The chickens, which have been feral on Kauai for years, actually don't represent much of a problem. The pigs and goats have become the destructive invaders. They don't intend to disrupt the island's plants, birds, and animals as they enjoy the balmy climate, but they do. The pigs root up native plants, which are unable to compete with imported weeds, and the goats consume the native plants. It's a story of destruction in progress, with rare and valuable plants disappearing. One extreme botanist, who rappels down cliffs to visit the last remaining plants of some species, bows his head in prayer when he finds yet another has disappeared since his last visit.

This information opened my eyes to the situation, but what caught my attention was the story of the large flightless ducks that were the island's largest fauna until they were hunted to extinction as food by Polynesian settlers one thousand or more years ago. DNA analysis suggests they were more closely related to dabbling ducks than geese, although their size, nine to sixteen pounds, is more in the range of geese. They had a tortoiselike beak and ate mostly leaves and ferns.

Considering Hawaiian birds led me to the Nene goose, the state bird and a critically endangered species. These charming geese are so sweet-tempered that they tame with a few handouts from humans. That's endearing, but it also makes them subject to becoming roadkill and attacks by local dogs.

After I returned, I began looking more closely at the history of Nene geese. A dedicated group of scientists and rangers are helping the species restore its numbers. Much has changed since the last comprehensive books were written about them in 1980 and 1999. Except for children's books, no books about the Nene remain in print. I felt called to write one.

My editor at Voyageur Press, Danielle Ibister, liked the idea but thought it too narrow for Voyageur's list. She suggested that I include information about the Nene in a book about the wider group of birds that people raise in small flocks. And thus this book got started.

I am grateful to Danielle for helping guide and direct the project. I am grateful to all who helped me along

The Nene goose of Hawaii helped inspire this book.
Shutterstock

the way. Craig Russell, president of the Society for the Preservation of Poultry Antiquities, read every word, walked me through technical details I didn't understand, and added history that I didn't know. He has been unfailing in his support. He knows more about poultry than anyone out there, and I am thrilled and grateful to call him my friend and colleague.

The American Poultry Association and the American Bantam Association shared their illustrations. As the umbrella organizations for poultry, they are indispensable.

Many people helped along the way: Lou Horton, premier waterfowl expert; Harvey Ussery, small-farming guru; Tom T. Walker, turkey and goose expert extraordinaire; Frank Bob Reese, advocate for traditional breeds and single-handed turkey breeder and marketer; G. Philip Bartz, longtime chicken judge and breeder; Barry Koffler, webmaster of www.feathersite.com; Patrick Sheehy, dedicated to Hookbill ducks; Bernd and Marie-Anne Krebs, devoted goose fanciers; Jeremy Trost, who gives so selflessly of his time and good humor; Mike Walters of Walters Hatchery; Donnis Headley, for the inspiration of her ability to turn a phrase relevant to poultry; and Carol Cook of Cook's Peacock Emporium, Dianna & Robert Nehring of Mouse Creek Feather Farm & Hickory Ridge, Bart Harrell of Grand Ridge, Florida, and Dennis Erdman of Erdman's Game Farm for their advice on peafowl. Mr. Harrell and Mr. Erdman are founding members of the United Peafowl Association. Myra Charleston, At Large Director of the American Emu Association, took time away from her mother's sickbed to patiently advise me on her subject. Joanne Rigutto shared her wonderful emu pictures with the readers. Craig Hopkins, Hopkins' Alternative Livestock, UPA, Inc. secretary, advised on several species and generously shared his knowledge and photos of rheas. J. D. Engle of Claude Moore Colonial Farm enthusiastically contributed photos and information about historic breeds at living history museums. Jeannette Ferguson, guinea fowl maven and author of *Gardening with Guineas*, advised me on their habits. Geoffrey R. Gardner, DVM, Lakeland Veterinary Hospital, swan veterinarian and expert, helpfully read the chapter on swans. Kermit Blackwood, pheasant expert, guided me on those species and other rare land fowl.

David Sweet, director of operations for Eurasian Feather Inc./Down Inc. was unstinting in giving his invaluable advice with cheerful willingness to inform me about his specialty. Tom Biebighauser advised me from his vast knowledge of building wetlands to create habitat. John Metzer of Metzer Farms patiently shared his knowledge and photos of waterfowl and game birds.

Merritt Clifton, editor of *Animal People*, the leading independent newspaper providing original investigative coverage of animal protection worldwide, is in a class by himself. He knows everything worth knowing about the relationship between animals and people and willingly offered his insight. Every time Merritt sends me information, I learn something.

I am grateful to all who helpfully tutored me along the way. Any errors that have crept in are my own.

I am grateful to NTBG Director of Education Namulau'ulu Tavana and Environmental Journalism Fellowship Course Facilitator JoAnn Valenti for this invaluable project. The experience I had there changed my life.

Each book is a journey for the writer, who needs all the support she can get along the way. My sisters in spirit of the Artist's Way group have buoyed me up more times than I can remember. My friend and legal adviser, Susan, was always there. Lunch and long walks with Lorraine helped me keep perspective.

There must be a special place in Heaven for spouses and children of writers. Gordon has stepped up to his role with good humor and endless affection. My daughter, Sara, formerly known as Nicole, remains my guiding light. I would never have stepped onto this path without her encouragement.

Ruth Ann Harnisch has continued her invaluable personal and professional support. She instantly offered financial support to help me take advantage of the NTBG fellowship that began this journey. She has given her no-nonsense advice every time I have asked for it. Neither of these books would have been possible without her help. Writing them wouldn't have been as much fun, either.

The Nene book remains to be written. I look forward to it.

OVERVIEW

The term *poultry* refers to all domestic fowl. In the context of this book, poultry is expanded beyond traditional fowl to include a few other bird species that are raised as livestock. *Fowl* encompasses chickens, ducks, geese, swans, guinea fowl, and some domesticated game birds, such as quail, pheasant, and peafowl. It's not a scientific term. Because people are also interested in raising, at the small end, pigeons and, at the large end, ostriches, emus, and rheas, those birds are included. This book will give you the basics necessary to get you started.

Scientifically, chickens, turkeys, guinea fowl, quail, pheasant, and peafowl belong to the order Galliformes, within the Aves class of birds. They are called land fowl, or gallinaceous birds. Many remain wild; others become tame relatively easily and have been domesticated for thousands of years. For example, wild turkeys are the ancestors of all domesticated turkeys. These birds form the foundation of the domestic poultry we enjoy today. Other land fowl, such as chukar partridges from Asia and guinea fowl from Africa, retain their wildness.

Gallinaceous birds typically nest on the ground. Their chicks walk and peck for food soon after hatching, characteristics called *nidifugous* (Latin for "fleeing the nest"). Most gallinaceous birds tend to be omnivorous, eating grains and plants as well as insects and other invertebrates, such as snails. Most species are social, preferring to live in flocks or mated pairs, but some are solitary. Land fowl have varied types of communication, from vocalizations to acoustic signaling to making whirring noises with their wings and tails. It's a lively world.

Ducks, geese, and swans belong to the order Anseriformes. They are called waterfowl, or anserine birds. They love the water, spending their lives in or near it, building nests in marshes or near lakes. Some even spend part

Poultry fit in well with other animals. These white Pekin, brown Khaki Campbell, and black Cayuga ducks and the cat show a mutual curiosity on a rural property in Maine. Poultry kept in small flocks are companionable yet independent. They are willing to forage for food on pasture or can be kept in confinement. Small bantams and pigeons have minimal space requirements. *Art Lindgren*

To a Waterfowl

By William Cullen Bryant

Whither, 'midst falling dew,
While glow the heavens with the last steps of day,
Far, through their rosy depths, dost thou pursue
Thy solitary way?

Vainly the fowler's eye
Might mark thy distant flight to do thee wrong,
As, darkly seen against the crimson sky,
Thy figure floats along.

Seek'st thou the plashy brink
Of weedy lake, or marge of river wide,
Or where the rocking billows rise and sink
On the chafed ocean side?

There is a Power whose care
Teaches thy way along that pathless coast,
The desert and illimitable air,
Lone wandering, but not lost.

All day thy wings have fann'd,
At that far height, the cold, thin atmosphere,
Yet stoop not, weary, to the welcome land,
Though the dark night is near.

And soon that toil shall end;
Soon shalt thou find a summer home and rest,
And scream among thy fellows; reeds shall bend
Soon o'er thy sheltered nest.

Thou'rt gone, the abyss of heaven
Hath swallowed up thy form; yet, on my heart,
Deeply hath sunk the lesson thou hast given,
And shall not soon depart.

He who, from zone to zone,
Guides through the boundless sky thy certain flight,
In the long way that I must tread alone,
Will lead my steps aright.

U.S. Fish and Wildlife Service

Poultry Language

Chicken

- "Chickens come home to roost" means that the consequences of ill-advised or immoral actions of the past are being felt.

Duck

- The duck test: "If it looks like a duck, walks like a duck, and quacks like a duck, it's probably a duck." This adage means that things are probably exactly what they appear to be. It is used in the context of arguments that reach to make excuses or find alternative explanations. It is attributed to James Whitcomb Riley, who wrote in one of his poems that "when I see a bird that walks like a duck and swims like a duck and quacks like a duck, I call that bird a duck."

- "Getting your ducks in a row" means being well organized, making all the appropriate preparations before embarking on a project.

Goose

- A *goose* is a person deficient in judgment and good sense.

- A *goose egg* is a big zero, especially when written as a numeral to indicate that no points have been scored.

- "What's sauce for the goose is sauce for the gander" means that what is suitable for a woman is suitable for a man. The origin of this phrase goes back to the seventeenth century.

Turkey

- *Turkey* is used pejoratively to mean a stupid or incompetent person.

- A *turkey* can also be a conspicuous failure, especially a failed theatrical production or movie.

- To *talk turkey* is to speak frankly and get down to the basic facts of a matter.

Peacock

- A *peacock* is a vain, strutting person.

Pigeon

- A *stool pigeon* is someone who betrays a jailhouse confidence, a person acting as a decoy or as an informer, especially one who is a spy for the police. It comes from the use of a pigeon as a decoy.

- *Pidge* is a term of affection.

---

of their lives on salt water. Their chicks are nidifugous, too, walking, swimming, and feeding themselves shortly after hatching. Primarily herbivores, anserine birds feed on plants, but some eat snails and other critters in their watery environment. Some waterfowl form large flocks and even nest together in large groups.

Pigeons are members of the family Columbidae, within the order Columbiforme. They are different and separate from other bird species. The pigeon species that have been domesticated have widely varied colors, feathers,

and behaviors that make them distinctive. Their small size makes them accessible for domestic husbandry.

Ostriches, emus, and rheas belong to separate families in the order Struthioniformes. The general term for these birds, ratite, comes from the Latin word for "raft"—a reference to the fact that that these birds have no keel on their sternum. A fowl has a keel-like prominence on its breastbone, whereas a ratite has a flat breastbone.

Ratites are flightless and lack the large breast muscles of flying birds. Their meat is on their legs. Ostriches have

The Arrow or Reeve's pheasant is native to Northeastern China. It inhabits steep, rugged terrain with rock escarpments and tall conifers. The Arrow pheasant's retrices, or tail feathers, grow up to six feet in length. These tail feathers help it fly much higher and faster than typical pheasants. It is classified as Vulnerable on the IUCN Red List, due to habitat loss and over-hunting it for food and those remarkable tail feathers. Private collectors continue to raise enough that it does not face extinction in captivity, but poor management practices have resulted in badly inbred stock. *Shutterstock*

Poultry make a good family project. Newly hatched chickens, ducks, geese, turkeys, and many game birds can be safely shipped via the U.S. post office on the first day of life without food or water. They continue to absorb the yolk for the first day or so of life. Giving the post office advance notice helps mail-order shipments succeed. *Shutterstock*

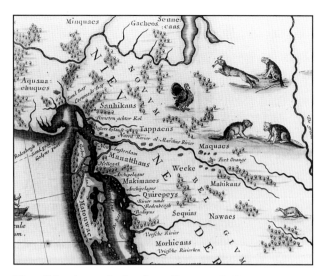

The wild turkey and other fauna illustrate Guiljelmus Blaeu's 1635 map of New Netherlands. European artists based their depictions on the bird they knew from Europe. The turkey was a distinctive American bird that represented local fauna. *Sabine Eiche*

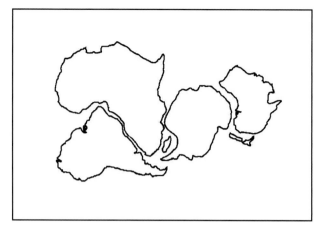

This map shows the present outlines of the continents that composed Gondwanaland, although they did not have their present shape then. Separation occurred over a long period of time. South America and Africa separated first at the southern end, gradually rotating away from each other over 60 million years. Antarctica has rotated 90 degrees clockwise and Australia 90 degrees counterclockwise with respect to each other. *Cracraft, 1973*

wings used in mating displays, although the birds themselves are far too large to fly. Emus have only vestigial wings. Rheas retain large wings that they use to run fast and take corners.

ORIGIN AND HISTORY

Birds developed from the reptile dinosaurs of the Triassic period, 195 million to 225 million years ago. Their forelimbs evolved into wings, their bones honeycombed to reduce weight, their scales became feathers, and they took to the air. Many poultry species still have scales on their legs. All birds lay eggs, as dinosaurs did. The oldest bird fossil is Archaeopteryx, some 150 million years old.

Life evolved in the tidal waters of the salty oceans, with water birds the first true birds, emerging in the warm Cretaceous period 135 million years ago. The ice of the Cenozoic era wiped out much life. As the continents broke apart from each other under the pressures of plate tectonics around 60 million years ago, Australia drifted away from the southern supercontinent of Gondwanaland. Birds west of the division, such as jungle fowl, are significantly different from birds east of it, such as emus. By 30 million years ago,

during the Oligocene epoch, a bird called *Cygnopterus affinis*, similar to the mute swan, lived in what is now Belgium. A similar bird, *Cygnavis senkenbergi*, left its bones behind as fossils in Germany, having lived 25 million years ago during the Miocene epoch.

Birds flourished in the warm Pliocene period, 1 million to 5 million years ago. Many different kinds of birds developed during that time, including the ancestors of ducks, swans, ostrich, and land fowl. Fossils similar to the mute swan suggest that mutes were indigenous to North America at that time. Although separated from the birds on other continents, the black swans of Australia are related to other Cygnus species.

By the time humans appeared 250,000 years ago, many fowl flew through the air, nested in trees, and ran on the ground. Birds and eggs must have played a significant role in the human diet. Swans and other birds are depicted in the cave art of Cro-Magnon humans, from 30,000 years ago. With animal domestication occurring about 10,000 years ago, humans began their partnership with animals. Fowl were among the first creatures domesticated. Domestication is documented to 5,000 years ago in China.

This Japanese woodblock print of children fighting their birds was created in 1772 by Edo artist Shigemasa Kitao. He is known for his naturalistic presentations of birds and flowers, as well as human subjects. It attests to cockfighting's history as a popular sport.
Library of Congress

Trumpeters cup their wings as they come in for a landing on water. They thrust their huge webbed feet forward to brake. Their air speed drops and they skim along the surface of the water until they come to a stop in a swimming position.

Flight has always drawn our eyes to birds in the heavens. We continue to study how they manage this wonder. Biologist Carroll Henderson explores refinements like the trumpeter swan's ability to land on water without making a splash in his book *Birds in Flight*.

POULTRY IN MODERN LIFE

Modern industrial poultry operations may house twenty thousand or more egg layers in a single structure. Laying hens are confined to the smallest possible cages, called battery cages, which are stacked on top of each other. The minimum requirements of the United Egg Producers call for 67 to 86 square inches per bird. Conditions are criticized as filthy and inhumane, with manure from cages above dropping onto birds beneath.

Cornish/Rock chicken crosses raised for meat are confined their entire lives to houses of forty thousand to one hundred thousand genetically nearly identical birds. Such conditions encourage the development of highly pathogenic organisms like H5N1 avian influenza, commonly known as the bird flu, which has fueled fears of becoming a pandemic.

Advances in breeding for commercial purposes have included a featherless chicken that was introduced in 2002. The horrified reaction of the public to this sad creature prompted its rapid withdrawal. The mindset of the commercial producer, who views such an innovation as progress, is vastly different from that of the small-flock owner, who views his flock with affection and admiration.

Five Freedoms

The Five Freedoms are the principles of livestock treatment distilled from discussions of humane animal care. They are used by animal organizations like the Animal Welfare Institute, which certifies farms for humane treatment of livestock. When in doubt about how to treat your poultry, return to these fundamentals.

The Five Freedoms are:
1. Freedom from hunger and thirst.
2. Freedom from discomfort.
3. Freedom from pain, injury, and disease.
4. Freedom to behave normally.
5. Freedom from fear and distress.

Disposing of the manure from such concentrated animal confinements poses substantial environmental problems. Rivers, streams, and coastal waters become contaminated by concentrated animal waste. Laying hens that are past their economic value, known as "spent layers," are euthanized and composted. Occasional reports of hens surviving that ordeal have attracted outraged attention to these practices.

The Cornish and Plymouth Rock breeds account for virtually all the commercial chickens and eggs produced today. Reliance on such limited genetics exposes the industry to disaster. The food pipeline bringing cheap chicken to the local grocery store is teetering on the edge of high fuel prices and environmental degradation. Public awareness of where our food comes from and how it gets to our tables is shrouded by what Ann Vileisis in *Kitchen Literacy* calls a "covenant of ignorance" between producers, manufacturers, and consumers. Unfortunately, protecting consumers from the unpleasant details of animal agriculture also shields producers from public scrutiny and regulation.

Consciousness about the conditions under which food animals are raised and the impact on rural communities has reached the mainstream. *The New York Times* editorialized about the subject in 2008, saying, "The so-called efficiency of industrial animal production is an illusion, made possible by cheap grain, cheap water and prison-like confinement systems. In short, animal husbandry has been turned into animal abuse . . . and [has] helped empty and impoverish rural America."

Cornish/Rock crosses are the dominant meat birds raised today. Poultry is an important segment of the American food production system. Over 9 billion broiler chickens are raised in the United States each year, nearly 50 billion pounds, valued at close to $9 billion.
U.S. Department of Agriculture

Intensive industrial facilities contain thousands of chickens to meet market demand for low-cost meat and eggs. The concentration of so many birds in confinement raises issues of air and ground water pollution and disease, for both human and bird. Relying on centralized livestock operations for food supplies makes the system vulnerable to major disruption if a single location is affected by infection or natural disaster. Recent food-borne illnesses have defied forensic tracing. *Howard F. Schwartz, Colorado State University*

The Pew Commission on Industrial Farm Animal Production funded a two-and-a-half-year study of industrial animal production and concluded that the "current industrial farm animal production system often poses unacceptable risks to public health, the environment and the welfare of the animals themselves."

The International Assessment of Agricultural Knowledge, Science and Technology for Development (IAASTD) issued a report from four hundred scientists arguing that the focus on industrial production goals has resulted in "a degraded and divided planet." It recommends changes to institutional, economic, and legal frameworks that protect and conserve natural resources like soils, water, forests, and biodiversity while meeting production needs. It calls for a major paradigm shift that would place strong focus on small-scale farming.

Small integrated farms produce far more, proportionately, than industrial operations. The 1992 U.S. agricultural census showed that the average four-acre farm grossed $7,424 per acre and netted $1,400 per acre. Farms larger than six thousand acres averaged a gross income of only $63 per acre, netting $12 per acre, according to an analysis done by Peter M. Rosset. There's a reason they call small-scale farming "sustainable."

Secure poultry buildings can be an attractive part of the landscaping. This duck house adds a bright splash of color. It takes advantage of the slope and keeps its occupants safe from predators, even in a forested setting. Their wading pool is located behind the building. *Art Lindgren*

Pekins and Khaki Campbells investigate their surroundings. Ducks enjoy foraging for small invertebrates such as insects and snails. They need enough water to keep their nostrils clean of dirt and food but do not require water for swimming. *Art Lindgren*

A female ostrich extends one wing to preen her feathers. Ostrich feathers are a significant product in demand, mostly for costumes. The most desirable feathers come from ostriches aged three to twelve years, but ostriches may continue producing feathers for harvesting for as long as thirty-five years. Their loose feathers do not have the web that makes other birds' harder, more compact feathers cling together. *Shutterstock*

Those who raise small flocks play a significant role in this changing economy. Currently, consumers make extensive use of all these varied birds. They are raised primarily as food, often delicacies for gourmet markets. Quail eggs represent just one sought-after tidbit. Pheasant makes a rare and special meal. Goose was once the center of the traditional holiday feast, although it has lost market share to the turkey. Turkeys better tolerate the crowded conditions required by the commercial industry. Geese now represent less than 1 percent of commercial poultry production.

Around 22 million ducks are raised annually in the United States. In 2006, about one thousand ostrich growers in the United States raised about one hundred thousand birds. Emus are raised in at least forty-three states by about ten thousand families, three thousand of whom live in Texas. The emu population is about a million. Rheas are the newest U.S. farm-raised ratite, but with more than fifteen thousand birds in the country, they represent the world's largest population of farmed rheas.

Specialty food markets beckon. Organic products are the fastest-growing sector of the food market, growing more than 20 percent every year since the 1990s. Asian and South American restaurants are outlets for quail eggs, while Indian restaurants are outlets for pheasant and quail table birds. As fuel costs increase, local food producers have an advantage

Guinea fowl are gaining popularity both as barnyard fowl and for meat and egg production. Although less willing to adapt to confined life, they can play an active role in pest management as they make their way around the barn and pastures. They can also be raised in confinement for meat production, a market segment that remains underdeveloped. *Shutterstock*

over commercial producers, who must transport food over long distances.

Local food has become so popular that a new word, *localvore*, has emerged. Like omnivores or herbivores, localvores are defined by what they eat—in this case, only locally grown food. Consumers are demanding more accountability from producers. They are willing to pay a premium to get good food. Rising fuel costs are changing the equation for corporate giants. Transporting food over long distances—the average quoted for grocery store food is 1,500 miles—levels the playing field between small and large producers. "Sooner or later, we're not going to be able to afford to haul food in from everywhere in the world," conservationist author Wendell Berry said in an interview in 2008. Food security takes on more urgency. Supporting local

food producers develops a safety net of food that will be available regardless of global fuel crises.

Food-borne illness is constantly in the news. The U.S. Department of Agriculture (USDA) frequently recalls meat and poultry from the shelves due to contamination. Large producers can contaminate huge quantities of food, which are rapidly shipped across large geographic areas before illness is detected and traced to the source. The possibility of food spoilage and contamination is ever present. In contrast, small producers are less likely to confront the highly pathogenic microorganisms that thrive in crowded animal confinement conditions. Their smaller reach makes tracing quicker and affects fewer people.

Small-flock owners are ideally suited to meet today's public and social challenges and market demands.

Slow Food USA

Slow Food USA is a nonprofit educational organization dedicated to supporting the food traditions of North America. It promotes good, clean, and fair food.

The organization is a resource poultry raisers and consumers can use. By focusing attention on traditional breeds, it encourages more farmers to raise these birds and consumers to seek them out. It offers ideas for cooking methods and recipes. Organized support will help traditional breeds flourish and find routes to delivering their unique contributions to our culture.

Its Ark of Taste project aims to revive rare regional foods. It includes nine heritage chicken breeds (Buckeye, Delaware, Dominique, Java, Jersey Giant, New Hampshire, "Old Type" Rhode Island Red, Plymouth Rock, and Wyandotte) and eight heritage turkey varieties (American Bronze, Black, Bourbon Red, Jersey Buff, Midget White, Narragansett, Royal Palm, and Slate).

BEYOND FOOD

Many birds are raised as game and released for hunting, including pheasants, quail, bobwhites, and partridges. Wild turkey programs that have relocated birds and improved habitats have resulted in increases from a 1973 low of 1.3 million birds to nearly 7 million birds across North America today.

Guinea fowl provide insect control and are also kept as exhibition birds. They make good table fowl, and their advocates enjoy their eggs, figuring two guinea eggs equal one large chicken egg.

Exhibition is important for all poultry. In many cases, exhibition fanciers prevent breeds from disappearing entirely.

 The American Poultry Association was established in 1873 and published the first *Standard of Excellence* in 1874. At that time, it represented both commercial and exhibition poultry interests. Today, it focuses primarily on exhibition. It is the central national authoritative organization promoting *Standard*-bred poultry, training judges, recognizing Master Exhibitors, and setting breed standards. *Copyright American Poultry Association*

 The American Bantam Association represents the interests of purebred bantam chickens and ducks. Bantams include miniature varieties of standard-size breeds, as well as breeds that have no large-fowl correlates, such as the Nankin and the Black Dutch. The latter are called true bantams. The ABA has its own *Standard*, describing 57 breeds, 85 plumage patterns, and more than 400 varieties of chickens and ducks. The ABA licenses judges and honors Master Exhibitors. *American Bantam Association*

 The Society for Preservation of Poultry Antiquities was founded to protect and preserve, for historical, educational, and recreational purposes and in the public interest, *Standard*-bred domesticated chickens, waterfowl, turkeys, and guineas. Since it was founded in 1967, it has acquired nonprofit 501c(3) status and actively advocates for traditional breeds. *The Society for Preservation of Poultry Antiquities*

 The American Poultry Association and the American Bantam Association work cooperatively on projects such as the APA-ABA Youth Program. Kids can start as young as five years of age to learn about poultry and participate in shows and other events. Young people with a foundation and love for *Standard*-bred poultry are in demand for leadership in the poultry industry. *APA-ABA Youth Poultry Program*

As with other livestock, poultry breeds that have been raised throughout history are being lost. The United Nations Food and Agriculture Organization (FAO) commissioned a report in 2007 on loss of livestock breeds. The Consultative Group on International Agricultural Research (CGIAR) worked with the International Livestock Research Institute (ILRI) to determine what breeds are being raised. As industrial breeds are adopted for their prodigious production, such as the ubiquitous Cornish/Rock, local breeds are lost. "Valuable breeds are disappearing at an alarming rate," said Carlos Seré, director general of ILRI. "In many cases we will not even know the true value of an existing breed until it's already gone."

The report recommends establishing gene banks to save genetic material. However, technical problems prevent preserving poultry eggs and sperm. Realistically, it's up to small-flock raisers to keep these breeds alive and vigorous.

"The international community is beginning to appreciate the seriousness of this loss of livestock genetic diversity," said Seré. "FAO is leading intergovernmental processes to better manage these resources."

The Global Crop Diversity Trust has established the Arctic Seed Vault, shown here with director Cary Fowler, to maintain a collection of crop seeds. The vault is located in the Arctic so that, in the event of a power failure, the invaluable seeds will be saved. The website describes it as "the ultimate safety net for the world's most valuable natural resource." Technical problems make keeping poultry reproductive materials impossible. Breeds must be kept as living birds to survive. *Global Crop Diversity Trust*

Livestock conservation organizations like the Society for Preservation of Poultry Antiquities and the American Livestock Breeds Conservancy encourage small-flock owners to keep historic breeds. Keeping a historic breed makes you part of livestock conservation. Each additional flock increases the genetic strength of these birds.

Endangered Land Fowl

Many land fowl species remain wild, but as with other wildlife, their existence is threatened by loss of habitat and overhunting. Many are already extinct throughout much of their former range. As conservationist Kermit Blackwood says, "These last few captive populations of wild species may well represent the last stock available to us. As such, they are treasures that must be maintained with a stewardship mindset."

Due to avian influenza and other infectious diseases, importation of birds between countries is highly regulated and so restricted as to be, for all practical purposes, impossible. Existing captive populations must be carefully husbanded, a challenge because they must be maintained as wild land fowl rather than domestic poultry.

"They are somewhat expensive but well worth the dedication of the serious hobbyist," says Blackwood. "Like domestic poultry conservationist stock, they are not suitable for beginners. However, they are ideal for experienced breeders who are prepared to move from domestic and ornamental game birds to the stewardship of invaluable wild stock. The steward of endangered stock must practice ethical husbandry practices, including keeping detailed records on bloodlines. Collaborating with other aviculturists in the conservation breeding of rare stock enhances the possibility of species survival."

CHICKENS

Chickens are the domesticated *Gallus* that has accompanied the advancement of human civilization. Chickens were one of the earliest animals domesticated, no doubt in many places over time. They were likely domesticated in Southeast Asia as early as 4000 BC. The jungle fowl that are their antecedents are so tempting they would certainly have attracted humans to catch and keep them.

Red jungle fowl still live in the wild in much of south and southeast Asia. This cock was photographed in Rajaji National Park in north-central India. Their crows echo through the dense forests as they have for thousands of years. They may be increasingly breeding with domestic chickens, which roam free at the park's outskirts. Jungle fowl traits typically dominate over domestic traits, meaning the chicks hatched from these breedings likely retain the flighty nature of jungle fowl and return to the jungle. The domestic genes they brought with them in turn remain in their progeny. *Tomas Condon*

Since the mists of prehistory, chickens have spread around the world. New evidence of domesticated chickens in South America adds strength to the idea that South Pacific Islanders and Polynesians made contact with the West Coast long before Columbus arrived from Europe on the East Coast.

Chickens were most likely originally kept as sporting birds, for fighting contests between game cocks. But they acquired religious and spiritual importance early in their journey with humans. Jungle fowl crow in the morning, which associated them, in human minds, with the sun. Spiritually, they chased away the dark spirits of the night. Hens lay eggs year-round and raise multiple chicks, with one rooster leading a flock of hens, making them natural symbols of fertility. The solicitous behavior a hen exhibits toward her chicks resonates with our feelings of maternal love.

Small in size and adaptable to almost any climate, chickens were Everyman's livestock. Even those who couldn't afford to keep large animals like horses and cattle could afford a small flock of chickens. Without requiring much attention from humans, they naturally raise their own replacements. With domestication came year-round egg-laying, which provided a food source as

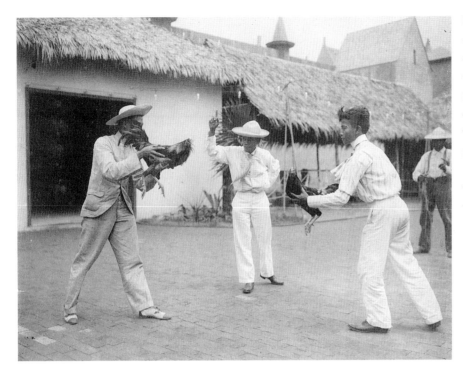

Fighting birds for sport, and the gambling with which the fights are associated, was an early use of chickens, as shown in this 1893 photograph by Frances Benjamin Johnston. Cockfighting continues today, although it is illegal in most parts of the developed world. Breeding muscle development into their birds was important to cock fighters, which led to the meat breeds such as Cornish.
Library of Congress

well as perpetuation of the flock. Chickens were suited not only to settled agrarian life, but they made productive companions on long journeys, whether crusade, war, or exploration. The fishermen of Southeast Asia raised a jungle fowl hybrid called the Ayam Bekasir that had an especially long, loud crow, allowing them to stay in contact with other fishermen. Chickens made their way into every corner of human life.

The captive populations of birds gave their keepers the opportunity to exercise selective breeding. Chickens reproduce relatively rapidly, faster than dogs or cattle. Differences in color, feather quality, size, comb, and other characteristics emerged rapidly, resulting in the development of distinct breeds. Such individual flock changes might have first occurred accidentally; for example, an enclosed flock, bred over a couple of years, may have become dominated by red tails or long sickle feathers. As people observed their animals over time, they learned that they could influence their flock by choosing those birds with desirable characteristics as breeders, and increase those characteristics in their flocks.

Such individuality at first might have had the advantage of simply distinguishing one farm's flock from another, reducing theft. Breeds became a way to manage the poultry on a farm. Aesthetics undoubtedly played a part, appealing to the owner's sense of beauty. Utility qualities, such as growing larger more quickly or laying more eggs, would certainly have been a source of pride as well as greater value and improved nutrition. Even one egg a day would have made a huge difference in a primitive diet.

Thus breeds were established from early on. By the thirteenth century when Marco Polo traveled from Italy to China, the small Silkie chickens he found there surprised him because they were so different from the smooth-feathered Continental breeds of his home.

Colonists in America took pride in their livestock, but organized breeding was less important than egg and meat production and animal survival. Traditional breeds like Dorkings and English Games accompanied them from England. Birds used for cockfighting, a popular sport, acquired the name Dung Hill Fowl. Later, Dung Hill Fowl described the mixed breeds that resulted from catch-as-catch-can barnyard breeding. "Hen fever" emerged as a craze in the mid-nineteenth century. Wealthy businessmen became caught up in the fad of breeding and exhibiting chickens. The arrival of the Java breed in the 1840s was a major event, and the

Silkie chickens have feathers more like hair than traditional chicken feathers. They require some special handling, because their feathers do not repel water. If they get soaked and cold, they can die. Their skin and bones are black, making them prized in Asia for medicinal qualities. Silkies are known as good setting hens and mothers. This Partridge color variety hen shows her chick how to forage for tasty bugs. *Shutterstock*

These drawings represent the ideal for Buckeye chickens in the 1915 APA *Standard of Perfection*. In its black and white edition, the APA continues to rely on the exceptional artwork of Arthur O. Schilling for most breeds. Franklane L. Sewell, Louis Paul Graham, and I. W. Burgess have also contributed artwork. The portraits were done from 1914 through 1952. *Copyright American Poultry Association*

Buff Cochin's arrival in England in 1845 fed the frenzy. The buff color was bred into nearly every chicken breed by the end of the century. The first American poultry show was held in 1849 in Boston, a tradition that continues to the present.

Eventually, the Civil War intervened, diverting attention from recreational pursuits. The violence of the war exterminated much livestock. After the war ended, those who worked in poultry production and exhibition renewed their efforts, refining and developing distinctive new American breeds, such as the Dominique and the Plymouth Rock, in all its color varieties. They organized the American Poultry Association and published the first *Standard of Excellence* in 1874. The *Standard* formalized judging standards for poultry shows. It remains the tool for breeding excellence for exhibitions. Now called the *Standard of Perfection*, it continues to evolve to meet changing conditions.

Poultry that was formerly kept by rural households and made locally available gradually became a consumer product. Chickens were transported to urban markets as populations migrated to cities. Various breeds gained their advocates as poultry producers competed for consumer dollars: Orpingtons, Rhode Island Reds, Cornish, Langshans, Plymouth Rocks of various color varieties, Wyandottes, Leghorns, and Houdans each held significant positions in the marketplace.

The growing poultry industry adapted technology to chicken raising after World War II. Traditional breeds lost ground to modern crosses developed to meet production goals. Where previously all chicks had been raised—males as meat birds and hens for eggs—separate breeds and varieties within breeds were now developed to supply separate markets for meat and eggs. Chicken breeding became a laboratory matter, to create the composite that would most quickly convert feed into meat and eggs. The result for meat is the modern Cornish/Rock cross, which reaches market size in six weeks; the

Traditional breeds were significant in the marketplace until twentieth-century hybrids and industrial agriculture methods replaced small-flock husbandry after World War II. This April 1910 issue of *Commercial Poultry* magazine features a White Crested Black Polish cock. Today, historic breeds are economically viable birds in the growing local food segment.

The Houdan is an old French breed. Its five toes indicate its relationship to English Dorkings. The White Houdan has been recognized for exhibition since 1914. It was developed from the original Mottled Houdan variety, one of the breeds included in the first *Standard* in 1874. Selective breeding eliminated the black from the plumage. *Robert Gibson*

Traditional breeds have been maintained both by fanciers as exhibition birds and by rural residents, who appreciate them for a broader spectrum of qualities, namely, their ability to raise their own chicks and perpetuate a flock, forage for their own food, and provide high-nitrogen manure to fertilize crops and gardens. Niche markets are developing for heritage-breed meat and eggs.

In many cases, these heritage chickens carry unique genes for beneficial traits like disease resistance. They are living examples of proud histories. Keeping a flock of traditional-breed birds allows your farm to conserve traits that may prove invaluable in the future.

FEED AND NUTRITION

Chicken nutrition has been exhaustively studied for the poultry industry. The goal for industrial operations is to convert feed into meat and eggs as quickly and as cheaply as possible. Your own flock of traditional-breed birds will take longer to mature. Commercial formulations are available as complete feeds. Purchase proprietary formulations from local feed companies, or create your own.

Chickens are omnivores. They eat both vegetable matter, such as greens and seeds, and animal matter, such as insects and small mammals. Both sources contain

result for eggs is the laying Leghorn. Cornish/Rock crosses suffer from congestive heart failure and skeletal deformities. Leg disorders often make it difficult for them to walk. Laying Leghorns reach the end of their economic lives in a year or two, at which time they are euthanized en masse and composted.

Along the way, the traditional breeds slipped from economic notice. The American Poultry Association, formerly the professional organization for producers as well as exhibitors, became less important than marketing organizations. The association is now exclusively concerned with exhibition fowl.

Chicken Tractors

Chicken tractors got their name because they perform the functions of a tractor in turning and fertilizing the soil. They are movable shelters that allow chickens to work over the soil floor, eating seeds and bugs, depositing manure, and working it into the ground with their scratching. As the resources in that area are used and the soil adequately worked, the structure is moved to a fresh area.

As with other chicken housing, tractors must be secure against predators and provide protection from any weather to which the chickens may be exposed. Your flock may have a separate chicken house and go into their tractor during good weather. Chicken tractors can also have enclosed and covered yards that provide protection during the day. Chicken tractors must be light for ease of movement and have wheels.

Chicken tractors have become popular, and many design ideas are available in books and on the Internet.

Shutterstock

the carbohydrates, oils, protein, vitamins, and minerals chickens require to be healthy and productive.

Chickens enjoy foraging for their own food. Consider whether your chickens will have foraging time in pasture or in a movable chicken tractor.

Freshness is important in feed. Commercial preparations are generally stable, but the nutrients inevitably decline over time. Feed is a perishable commodity. You can more easily monitor the freshness of feed that is made closer to home. Keep only as much feed as your birds will consume in a month or less. Some flock owners taste their feed to check for sour or rancid taste. Spoiled feed can kill chickens. Store it in a cool, dry place in a secure container. Keep bugs and mice out.

Formulated chicken feed comes as mash, crumble, or pellets. The crumbled and pelleted forms provide more uniform nutrition than mash. The fine grind of mash allows heavier ingredients to migrate to the bottom. Follow feeding directions. Most commercial preparations are designed to be fed freely. Limiting feeding to once or twice a day may result in the chickens not getting adequate nutrition.

Commercial formulated feeds provide complete nutrition, unless their tags specify otherwise. Feeding scratch grains and kitchen trim may not be recommended. They enjoy it so much, I can't imagine having chickens and not giving them interesting things to eat. Offer greens and other tasty treats as a supplement to commercial feed formulations.

Eating is an important part of a chicken's day. Scratching and pecking are natural behaviors that occupy most of their waking hours. If they are not on pasture where they can enjoy the results of their scratching, provide them with interesting things in their enclosure to attract their attention: hang dried vegetation from their fence, or give them garden clippings and kitchen trim. Without interesting things in their environment to forage, bored chickens will begin to peck at each other, which can lead to cannibalism. Directing their natural behaviors at appropriate targets can head off these problems.

Commercial feeds are tagged with a nutritional analysis of the contents. Protein is the most significant element. Growing chicks and laying hens need more protein than roosters and hens who aren't laying. Chick starter is 20

Broodiness in hens is a crucial quality for perpetuating a flock. Artificial incubators do an excellent job, too, but mother hens do it free and will not cool off if the power fails. This trait has been bred out of many breeds because broody hens produce fewer eggs than hens who stop laying to raise a family. For small-flock owners, broodiness is a useful quality. Historic breed conservation focuses on preserving useful traits such as broodiness, egg quality, and fertility that go beyond physical exhibition qualities. A mother hen caring for her chicks is an inspiring sight. *Shutterstock*

to 22 percent protein, broiler feed is 20 to 22 percent protein, and layer feed is 16 to 18 percent protein. More is not better. Excess protein in the diet can damage kidneys and cause skeletal problems. Give your birds feed that is appropriate to their needs.

Chickens on pasture will need supplemental feed. Chickens are generally self-regulating about their nutritional needs, but they may overeat commercial feeds in an attempt to supply deficiencies. Fat chickens do not lay well. If the fat pad between your hen's legs is cushy,

she's overweight. Consider more exercise or limit carbohydrates like scratch feed.

Most chick starter is medicated with a coccidiostat. Coccidia are protozoan parasites. Their eggs are commonly found in the soil of a chicken pen, so most chickens are exposed at some time. Coccidiosis can devastate chicks. The low dose of medication in feed allows them to experience some infection and acquire immunity. This kind of medication does not risk developing antibiotic-resistant infections. It is short term in the life of the

Vitamins

Nutrition research identifies thirteen vitamins required by poultry. They fall into two groups: fat-soluble and water-soluble. The fat-soluble group includes vitamins A, D3, E, and K. The water-soluble vitamins are the B-complex vitamins: thiamin, riboflavin, nicotinic acid, folic acid, biotin, pantothenic acid, pyridoxine, vitamin B12, and choline.

All these vitamins are essential for poultry and human life. Vitamin A affects the health and proper functioning of the skin and lining of the digestive, reproductive, and respiratory tracts. Vitamin D3 has an important role in bone formation and the metabolism of calcium and phosphorus. The B vitamins are involved in energy metabolism and the metabolism of many other nutrients. Brewer's yeast, or nutritional yeast, is a good source of B vitamins. It may be added to chicks' water in the thirteenth and fourteenth weeks of age to ensure adequate absorption of B vitamins and to help the chicks avoid leg problems that might otherwise develop at that time.

Eggs supply these vitamins to the developing embryo. That makes them a good animal source of vitamins for humans.

chicks, so it does not enter the human food chain. It's worth feeding to protect your chicks.

If you do not want any antibiotics in chick feed, ask your feed store to order antibiotic-free feed for you, or make your own.

Layer feed includes calcium, one of the significant differences between it and other feeds. Hens need calcium to produce healthy eggshells, which are made of calcium carbonate. Hens on pasture require a dish of calcium source such as oyster shell to meet their needs. Chicks should not get layer feed, as the extra calcium would interfere with their development.

Clean water must always be available. Lack of water will hurt your flock faster than any imbalance in their feed. Their bodies are 55 to 75 percent water, and eggs are 65 percent water. Chickens drink about twice as much water by weight as they eat. They breathe out moisture. Chickens have no sweat glands, so the evaporative heat loss in the air sacs of the lungs is the single most important way they cool themselves.

Chicken digestion depends on grinding the food into small pieces in the crop. They have no teeth, as conveyed in the expression "scarce as hens' teeth." Small sharp-edged sand and other inorganic material in the crop performs that function, allowing the bird to extract nutrition from the food. Birds on pasture will find their own. Birds that are confined should have a dish of grit available to them at all times.

HEALTH MANAGEMENT

Good nutrition, fresh air, and clean living conditions are the best defense against livestock diseases. Chickens can and do get sick and infested with parasites despite the best care, though.

Staphylococcus **bacteria** are common germs that can infect a cut locally. The infection can become systemic, meaning that it spreads throughout the body. Make sure the coop has no sharp edges on which chickens can cut themselves. Treat wounds with triple antibiotic ointment. If a chicken develops swollen joints or footpads or has a sore that oozes yellow pus, treat with erythromycin from the vet or feed store. *Escherichia coli* in dirty litter can infect chickens, causing diarrhea. Mycin antibiotics also treat that condition.

Avian Mycobacteriosis, or **tuberculosis**, causes nodules (called granulomas) to grow in internal organs. Infected chickens decline and eventually die. When symptoms occur in small flocks, birds are usually tested and infected individuals separated for treatment. Some commercial operators destroy the entire flock and disinfect the premises. The *Mycobacterium avium* is difficult to eradicate.

Aspergillosis, better known as **brooder pneumonia**, is a fungal disease caused by organisms that grow in contaminated feed and wet litter. Birds don't transmit it to each other. Once they develop symptoms, they are likely

A Light Brahma enjoys a dust bath. Chickens maintain their feathers by taking dust baths, which work dust particles into their feathers. They will find a spot on the ground with loose, dry dirt, or you can provide sand or peat in their yard. The sand grains remove oxidized feather oils, which the chicken replaces with oil from the uropygial gland at the base of the tail. Fresh oils keep feathers in good, fluffy condition. Dust baths also reduce external parasites and help chickens maintain cooler body temperature. *Shutterstock*

to die. There is no treatment. Aspergillus is a common fungus, and most birds do not get sick unless there are overwhelming amounts of the fungus. Clean and dry living conditions help flocks to prevent contamination.

Marek's disease is a contagious herpes viral disease in chickens that causes infected birds to grow tumors. Turkeys also occasionally contract it. Tumors grow in vital organs and the nervous system, eventually killing the chicken. All chickens are considered at risk. Chicks can be vaccinated on the eighteenth day of incubation or in the first days of life. Adult chickens that have not been previously vaccinated can be, and some farmers advise annual boosters for adult flocks. Make sure all the chickens you acquire are immunized, or immunize them yourself. An alternative is to obtain Marek's-free stock and keep a closed Marek's-free flock.

Infectious coryza, or **roup,** is like a cold but can be very severe. Chickens can develop secondary infections of the mouth and eyes that persist for months. Vaccines are available for farms that have a history of coryza, but they must be specific to the variant of the disease infecting the area. Early antibiotic treatment with erythromycin and oxytetracycline can help.

HOUSING

Chickens need secure housing. Although they can survive being confined indoors, they do much better when they have a yard to scratch in. After you've addressed the basic requirements, their housing is limited only by your circumstances and imagination.

As domestic birds, chickens are subject to predation. Their housing needs to be strong enough to resist the efforts of other animals trying to get at them. That includes the floor. Chicken wire or metal flashing buried around the perimeter will foil digging predators. A concrete floor can be hosed down, an advantage for cleaning. A deep-litter system is most successful on a dirt floor, to allow moisture to move through the litter and naturally occurring microbes to do their work. Deep-litter management on a concrete floor may require more frequent cleaning, and the manure may need to complete its composting process on an outdoor compost pile before being applied to plants.

Deep-litter management recruits the chickens' natural scratching and pecking behaviors into a balanced system of fibrous litter material. It creates a compost system of decomposing vegetative material, such as leaves, wood chips, or chopped corn cobs mixed with chicken manure, which is high in nitrogen. The composting mixture

Chickens are happy in humble surroundings but will adapt well to a chicken palace like this one. It was included on Raleigh, North Carolina's Tour d'Coop. Chicken housing should suit your budget as well as your landscaping. *Rick Bennett*

becomes a home for microorganisms, insect larvae, and other invertebrates that eventually turns into a desirable garden fertilizer. Cleaning is required only once or twice annually.

Chickens are roosting birds. Roosts should be round and smooth to avoid cuts to the birds. Having roosts all at one level avoids chickens jockeying over which one gets the top roost.

Chickens need protection from weather extremes. They are resilient and adapt to all climates well, although they need shade in hot climates and shelter from wind, rain, and snow in cold ones. While protecting them from the elements, their shelter also needs to be well ventilated. They need fresh air.

The site should be well drained. Swampy land can foster development of disease organisms like aspergillus. In wet areas, the chicken house can be constructed on blocks that raise it above the wet ground.

Chickens' toes curl naturally around wooden dowel perches. For large fowl, perches should be sixteen to eighteen inches apart, less for bantams. Square lumber two-by-twos with the edges rounded off and sanded can also be used. The natural variation of tree branches may suit your flock. *Shutterstock*

Space

Overcrowding causes many problems in raising chickens, from them pecking each other to outright cannibalism. Give your flock as much space as possible. Adjust the number of birds you have to the space available. Chickens should have at least three square feet per bird inside the henhouse. They should have at least five square feet per bird outside. More, in this case, is better.

Shutterstock

The Dorking is a foundation breed, meaning it is a historic breed from which modern composite breeds are bred. Dorkings are distinctive for having five toes. They can be identified in artwork from the Roman Empire era. They were established in the American colonies, but their long history in England keeps them classified as English for exhibition purposes. They are a dual-purpose breed, producing plenty of eggs as well as a meaty carcass for the table. *Shutterstock*

BREEDS

Chicken breeds are separated into foundation breeds and composite breeds. Foundation breeds are historic breeds from which modern composites were developed. Theoretically, if a composite breed like the Chantecler were to disappear, it would be possible for breeders to re-create it. However, the genetics of foundation breeds like the Java would be lost forever, as with the extinction of a species.

Foundation breeds are Java, Cochin, Langshan, Dorking, Hamburg, Campine and Braekel, Lakenvelder, Polish, Leghorn, Spanish, Minorca, Andalusian, Old English Game, Malay, Shamo, Sumatra, Phoenix, Aseel, Naked Neck, and Araucana.

More than fifty large-fowl breeds and more than sixty bantam breeds are classified by the American Poultry Association for exhibition purposes. They are divided into categories based on origin: American, Asiatic,

Selective breeding has produced a wide variety of chicken breeds, such as this White Crested Polish. Beneath the crest of feathers, the skull has a bony knob. Other crested breeds include Old French breeds such as the Crevecoeur and Houdan. *Fred Anderson*

Dominiques are a composite breed considered the first American breed, despite the French sound of the name. Many Americans knew them as Dominickers. The USDA recognized them in the Poultry chapter of the first USDA Yearbook of Agriculture in 1862. They are good egg layers, often laying throughout the dark days of winter. The hens retain broodiness and are good mothers. *Bryan K. Oliver, Dominique Club of America*

Frizzle feathers are caused by a specific gene and can occur in any breed. They were recognized in the first *Standard of Excellence* in 1874, and Charles Darwin remarked on them. Slim feathers curl better than broad ones. Despite their frivolous appearance, Frizzles can be as productive as their normally feathered cohorts. Except for the frizzled feathers, they must meet all other conformation and color standards for their breed in exhibition. *Shutterstock*

English, Mediterranean, Continental, and the catchall category of All Other Standard Breeds. The *Standard of Perfection* contains precise definitions of each breed and color variety against which birds are judged at exhibitions.

Breeds are defined by body shape and type, which are distinct to each breed. The definition of body shape includes dimensions and body proportions. The Cornish, for instance, are very broad across the shoulders compared to other breeds. Type takes precedence over color in selecting breeding birds. Poultry breeders abide by the maxim "Build the barn before you paint it."

Size and weight are significant breed characteristics. Rare breeds that have few infusions of new birds often lose size over years of breeding. At the same time, each breed has an optimum size that needs to be maintained in breeding operations.

Feather quality differs among breeds. Hard-feathered birds include Games and Orientals. Their firm feathers are held close to the body, very different from soft-feathered breeds like the Cochin. Silkies are unique in having feathers that resemble hair. They require special care, since their fine feathers are not as water-resistant as other chickens' and can become sodden if exposed to wet conditions.

Most breeds come in multiple color varieties. For example, ten color varieties of Leghorns are recognized, and breeders raise many more that are not recognized by the *Standard*. Color requirements are specified in the *Standard* for showing purposes. The primary, secondary, and main tail feathers may be different in color from the undercolor. Brassiness is a serious defect in color. Feathers may be solid in color or have penciling or laced markings. The Delaware's color pattern is its most distinguishing characteristic and important to its value in breeding for sex-linked chicks.

The skin on the legs, feet, face, and earlobes of a chicken may be black, red, slate, grayish-blue, or white.

Combs vary by breed, and some breeds have recognized varieties that differ by comb. Rose Comb and Single Comb varieties of Leghorns, Minorcas, and Anconas (all Mediterranean breeds) are recognized. Some breeds, such as the La Fleche and the Sicilian Buttercup, have unusual combs that are part of their charm as well as required elements of the *Standard*.

The Golden Laced Wyandotte *(above left)* was the first Golden Laced variety recognized by the APA *Standard*, in 1888. Buff Laced Polish *(above right)* were recognized five years earlier. Lacing refers to the contrasting color border around the edge of each feather. *Shutterstock (above left), Fred Anderson (above right)*

Heritage Definition

Frank Reese of Good Shepherd Turkey Ranch in Kansas, where he also raises chickens and ducks, has worked with individuals and organizations to devise a marketing definition for heritage chickens and eggs. In its abbreviated form:

A heritage egg can only be produced by an American Poultry Association Standard breed. A heritage chicken is hatched from a heritage egg sired by a Standard breed established prior to the mid-twentieth century, is slow growing, and is naturally mated with a long productive life.

This definition can help small-flock owners to distinguish their products from industrial ones and find a successful market for them.

SPPA Critical List

The Society for the Preservation of Poultry Antiquities, established in 1967, works to preserve historic or endangered poultry breeds. An "old" breed is defined as one developed prior to 1850, while a "rare" breed is defined as one with fewer than one thousand individuals in less than fifty flocks. While some breeds are plentiful, color varieties within a breed may have become rare and so are included.

Breed	Old	Rare	Varieties
American Game Bantam	X	X	All varieties
Ancona	X	X	Rose Comb
Andalusian	X		
Araucana	X	X	All varieties
Aseel		X	All varieties
Barnevelder	X	X	All varieties
Belgian Bearded d'Anvers	X		
Belgian Bearded d'Uccle	X		
Booted Bantam	X	X	All varieties
Brahma	X		
Buckeye	X	X	
Buttercup	X		
Campine	X	X	Both varieties
Catalana		X	
Chantecler		X	White and Partridge
Cochin	X	X	Silver Laced, Golden Laced, Blue, Brown
Cornish	X	X	All varieties
Crevecoeur	X	X	
Cubalaya	X	X	All varieties
Dominique	X	X	
Dorking	X	X	All varieties
Dutch Bantam		X	
Faverolle		X	Salmon and White
Fayoumi	X	X	
Jersey Blue	X	X	
Frizzle	X	X	All varieties
Hamburg	X	X	Golden Spangled, Golden Penciled, Silver Penciled, Black, White
Hollands		X	Barred and White
Houdan	X	X	White and Mottled
Iowa Blues		X	
Indian Games	X	X	
Japanese Bantams	X	X	Black, Black Breasted Red,

30

Breed	Old	Rare	Varieties
			Blue, Brown Red, Buff, Gray, Mottled, Self Blue, Silver Duckwing, Silver Laced, Wheaten
Java	X	X	All varieties
Jungle Fowl	X	X	All varieties
Kraienkoppe	X	X	
LaFleche	X	X	
Lakenvelder	X	X	
Lamona		X	
Langshan	X	X	White, Blue
Leghorn		X	
Long Crowers	X	X	All varieties
Malay	X	X	All varieties
Manx Rumpie	X	X	All varieties
Marans	X		
Minorca	X	X	Single Comb and Rose Comb White, RC Black, SC Buff
Modern Game	X	X	All varieties
Naked Neck		X	
Nankin Bantam	X	X	
Old English	X	X	All varieties
Orloff	X	X	All varieties
Orpington	X		
Persian Rumpless	X	X	All varieties
Phoenix	X	X	All varieties
Plymouth Rock	X	X	Buff, Silver Penciled, Partridge, Columbian, Blue
Polish		X	
Pyncheon Bantam	X	X	
Redcap	X	X	
Rhode Island Whites		X	
Rosecomb Bantam	X	X	All varieties, except Black
Rumpless	X	X	
Sebright Bantam	X		
Shamo	X	X	All varieties
Silkie Bantam		X	
Spanish	X		
Spitzhauben		X	

Breed	Old	Rare	Varieties
Sultan	X	X	
Sumatra	X		
Sussex	X	X	Red and Light
Thuringer	X	X	
Wyandotte	X	X	Golden Laced, Black, Buff, Partridge, Silver Penciled, Blue
Welsummer		X	
Yokohama	X	X	

In addition to these breeds, we would like to add the following based upon these criteria:
- Breed not listed either in the APA or ABA Standard.
- Breed must be found within the United States, U.S. Territories, or Canada.
- Breed must have historical references prior to 1900.

Many of these breeds are found mainly within the ethnic communities that first brought these birds to the United States and Canada (i.e., the Vietnamese GaNoi). Some of the game breeds listed are plentiful within the ethnic communities of origin but are otherwise difficult to obtain—thus, their rarity. Breeders are urged to use the breed standard from the country of origin, should one exist.

Breed	Country of Origin	Status
GaNoi or GaDon	Vietnam	Old and Rare
Jaerhon	Norway	Old, but carried by several hatcheries
Ko Shamo	Japan	Old and Rare, a true bantam
Malgache	Madagascar	Old and Rare
Minohiki	Japan	Old and Rare
Penedesenca	Spain	Old and Rare
Spanish Game	Spain	Old, but not Rare
Tomaru	Japan	Old and Rare

Buckeyes are increasing in popularity among small-flock owners for their large meaty size, good egg production, and beauty. Some Buckeye breeders prefer to select birds for egg production, while others prefer to retain the broody characteristic.
Bryan K. Oliver

PRODUCTS

Chickens are raised primarily for meat and eggs. Small producers can sell their products at farmers' markets and from their farms. Check local and state laws for regulations.

Traditional breeds are gaining notice in the marketplace for their superior taste and gourmet appeal. Organic poultry sales quadrupled between 2004 and 2006. The *Nutrition Business Journal* estimates that sales will continue to grow at 23 to 37 percent a year through 2010.

Barred Rocks, Buckeyes, Cornish, Jersey Giants, and New Hampshires are all good traditional breed choices for niche marketing. Local markets may be willing to partner with you in selling your product. Food brokers like Heritage Foods USA are selling table-ready Dark Cornish chicken. Local breeds that are not recognized by the *Standard* could become popular local products.

Research your market and make connections with brokers, restaurants, and retail outlets. Develop relationships to understand what your market needs, and tailor your products to meet them. The price premium (the higher price consumers are willing to pay for specialty products) and demand are increasing. Supply has not yet met the demand.

As part of the effort to keep traditional chicken breeds alive, Frank Reese started the Standard-Bred Poultry Institute. His goal is to turn his 160-acre farm in Kansas into a learning and resource center for heritage poultry breeds.

The future is bright for poultry products from traditional breeds. Whether you raise poultry solely for your own use or for income, many see increasing demand for these products.

CHAPTER 3

DUCKS

D ucks are waterfowl, so although it isn't required, they do better with some water to splash around in. If you have a pond on your property, ducks will use it happily. If you don't, most will manage without it, so long as they have a steady supply of drinking water. Some do require swimming water for mating.

Duck eggs are in demand, as are duck feathers and down. Their manure is a valuable fertilizer. Ducks thrive on kitchen and garden leftovers. They grow well on a wide and varied diet. They enjoy foraging for themselves in ponds, effectively reducing unwanted water plants, weeds, insects, and other pests like snails and mosquito larvae.

Ducks are beautiful, personable, and easygoing. You will be pleased that you have chosen to raise them.

HISTORY AND CULTURE

Ducks may have been domesticated even before chickens, as early as four thousand years ago in China. Pottery models of ducks and geese date back to 2500 BC.

The Indian Runner Duck traces its name back to ancient Bombay. The placement of its feet far back on its torso gives it its unusual upright stance. It can walk with a quick, straight step, unlike the waddle of other ducks. Its ability to cover ground on its feet would have been considered an advantage by Malaysian and other Southeast Asian flock owners who herded the ducks out to the rice paddies during the day—where they cleaned up snails, grain, and weeds—and back at night to shelter.

Ancient Romans and Greeks may have captured wild ducks and fattened them, or may have kept domesticated flocks. Marcus Terentius Varro, a Roman scholar and agricultural writer, left the first written record of raising ducks in 37 BC. Columella, who became a farmer after his military career in the first century AD, wrote advice about raising ducks in *De Re Rustica*, his twelve-volume tome on agriculture.

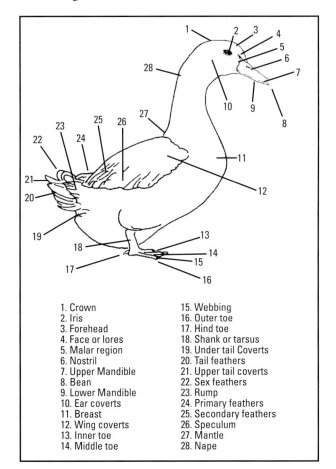

1. Crown
2. Iris
3. Forehead
4. Face or lores
5. Malar region
6. Nostril
7. Upper Mandible
8. Bean
9. Lower Mandible
10. Ear coverts
11. Breast
12. Wing coverts
13. Inner toe
14. Middle toe
15. Webbing
16. Outer toe
17. Hind toe
18. Shank or tarsus
19. Under tail Coverts
20. Tail feathers
21. Upper tail coverts
22. Sex feathers
23. Rump
24. Primary feathers
25. Secondary feathers
26. Speculum
27. Mantle
28. Nape

Copyright American Poultry Association

A children's wading pool can be easily cleaned to provide fresh water for ducks. Aylesbury ducks are a historic breed that has become rare. Their strong bodies and vigorous constitutions make them an excellent choice for small-flock owners.
Courtesy of author

Ducks Need Water

Ducks need fresh water for cleaning as well as drinking. It must be deep enough for them to dunk their heads completely. Otherwise, their nostrils can get clogged, leading to eye infections. It's ideal for them to have a place to get into the water, take a bath, and splash around to clean their feathers and drown parasites.

A large livestock watering trough can make a satisfactory waterhole for ducks. They will need a ramp to get into and out of it. Unless it's in a place where mud and holes are not a problem, wire-mesh platforms around the trough may be needed to protect the ground. During the winter, a five-gallon tub can be placed over a larger catch basin to avoid soaking the ground and making a muddy mess. The catch basin can drain into a line outside the house. Deep litter systems do not work when the litter gets soaked.

Short-legged ducks like Rouens and Aylesburys require water deep enough to swim in for good fertility. Males are often unable to mount females successfully on land.

Harvey Ussery

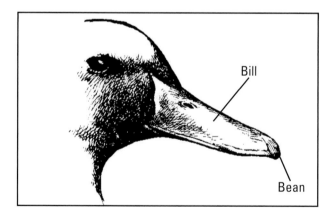

Copyright American Poultry Association

Copyright American Poultry Association

Egypt has always served as a flyway for migratory waterfowl traveling between Africa and Eurasia. Ducks and geese played an important role in the culture and food in the civilization of the pharaohs. Hundreds of thousands of ducks and geese visited the Nile, where many were trapped in hexagonal clap-nets.

Muscovy ducks were domesticated in Mexico and South America, despite their name. They may have acquired their name from the Muscovite Company, a trading company that took its name from Moscow, the center of Russian power, which began importing the ducks to Europe around 1550. They are non-migratory.

Ducks didn't flourish to the same extent as chickens and geese in medieval Europe, although the Mallard was probably domesticated in France as early as the tenth century. Ducks became common by the sixteenth century but did not show the increase in size associated with domestication until the eighteenth century. By then, distinct varieties were being recorded. Melchior d'Hondecoeter, the Dutch baroque–era (1636–1695) painter, featured Hookbill ducks in his piece *Ducks in a Landscape.*

The Muscovy became a popular farm duck by the 1840s. Pekin ducks, developed from domesticated Mallards in China, arrived in New York in 1873. That group of less than ten birds caught public attention, and the breed was included in the first *Standard of Excellence* in 1874. Aylesbury, Rouen, White Muscovy, Cayuga, White Crested, Gray and White Calls, and Black East Indies ducks were also recognized in that initial publication.

The duck's long association with humans has given cultures ample opportunity to attach many symbolic meanings to them. In China, Mandarin ducks are often included at weddings as symbols of wedded bliss and fidelity. Native American symbolism associates ducks with emotional comfort and protection.

British psychologist Richard Wiseman conducted a search for the world's funniest joke in 2002 and found that ducks were most often used as funny animals. Donald and Daffy lead the list of many cartoon characters that have made millions laugh. The University of Oregon is the only school licensed under a special agreement with the Walt Disney Company to use Donald Duck as its Fighting Duck mascot. Other athletic teams

The Madison Mallards is an appropriate name for a summer collegiate baseball team in the Northwoods League. The duck theme continues with their home park, Warner Park, known to fans as the Duck Pond. The logo and mascot add a playful spirit to team events. Fans appreciate it: The Mallards are the most popular team in the league, averaging over six thousand fans in attendance at games.

identify with ducks, such as the Madison, Wisconsin, Mallards, a summer collegiate baseball team.

Jemima Puddle-Duck and her sister-in-law, Mrs. Rebeccah Puddle-Duck, are beloved characters lovingly drawn by Beatrix Potter in *The Tale of Tom Kitten* and *The Tale of Jemima Puddle-Duck.* Ducks are everywhere in children's literature.

DUCKS IN AMERICAN DIET

Pekin ducks imported from China in the 1880s soon became established on Long Island, where they gained culinary respect. Table-ready ducks are often called Long Island Ducklings. Nearly all the ducks consumed in America today are Pekin ducks raised in the Midwest.

Americans consume an average of less than half a pound of duck a year. In the late nineteenth century, ducks were as popular as chicken but have since declined in favor of chicken and turkey. Duck eggs are available, but their popularity is limited mainly to Asian consumers. *Balut,* duck eggs incubated for sixteen days, are boiled and considered a delicacy among Filipino and Vietnamese ethnic communities. Salted duck eggs are also popular.

Some people who are allergic to chicken eggs may not be allergic to duck eggs. Duck egg yolks are higher in fat than chicken eggs, and the white is higher in protein. They substitute one-for-one for chicken eggs in cooking and baking. The higher protein content of the whites makes them whip up higher, making cakes lighter.

PRODUCTS

Meat is the most important product from ducks. Ducks can be butchered at home or by a custom processor. You should butcher them before they begin a molt to avoid encountering pinfeathers in the skin. Most ducks are fully feathered between seven and ten weeks of age, Muscovies between fourteen and sixteen weeks. If you have not butchered them at that time, wait until the feathers grow back, about six to ten weeks. Some people avoid the problem by skinning the carcass.

Hang the birds upside down or use a killing cone. Allow them to bleed out completely to avoid discolored meat.

Birds may be sold whole, halved, or as boneless breast meat. Confit is a preparation that preserves meat cooked

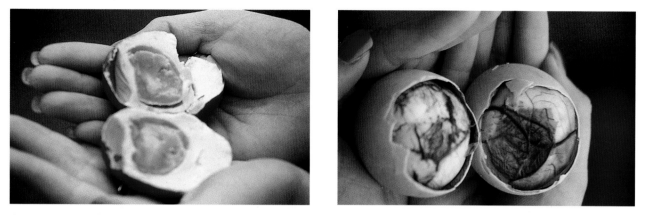

Salted duck eggs (*left*) are a Chinese and Filipino food item made by soaking the eggs in brine or wrapping them in salted paste. They cure for about four weeks and are then boiled or steamed. Balut (*right*) are a popular Vietnamese and Filipino food. Fertilized eggs are incubated for sixteen to twenty-one days, and then eaten. They are usually boiled but may be fried. Research local markets to identify products that can produce additional revenue for your operation. *Metzer Farms*

in its own fat. Although now a gourmet item, historically it was a way of preserving meat without refrigeration.

Feathers are also a valuable product. Waterfowl feathers are naturally curved, trapping air and providing better insulation than any synthetic product. Waterfowl also have down, the soft plumage closest to the breast and body. "Despite the considerable effort of many to develop a synthetic insulation that rivals down, nothing has been invented to this day that has even come close to providing the insulation, filling, and enduring quality of what nature created in both ducks and geese," says David Sweet, director of operations for Eurasia Feather Inc./Down Inc.

Duck down is less sought after than goose down but still makes a warm filling for bedding, clothes, and cushions. The quality of down is affected by the age of the bird, its lifetime nutrition, and its living conditions. Older, mature birds produce better feathers and down. Good nutrition and clean living conditions also help produce better products.

Feathers can be picked from the bird directly after it is killed, or the carcass can be scalded first. Scalding melts the fat that holds the feather in the skin.

Eggs should be gathered daily. Refrigerate them if they are to be eaten. Eggs will keep as long as six weeks, but of course fresher is better. Hatching eggs should be kept cool but not refrigerated—around 60 to 62 degrees Fahrenheit. Hatching eggs should be set within fourteen days. Wash them only if they are soiled, with a cool solution of vinegar and water, to avoid opening pores and driving any possible infection into the egg.

Commercial operations use every part of the duck. Tongues and feet are exported to China, where they are delicacies, and the bills are ground up for homeopathic medicines. Mink farms purchase the offal as food. One of the tenets of sustainability is finding ways to make use of all products.

FEED AND NUTRITION

Commercial formulations for waterfowl provide the nutrition to meet the needs of growing ducklings and goslings. If duck feeds are not available, ducklings can be fed chick starter.

Ducklings need more niacin than chicks, which can be added to the diet by adding brewers' yeast. Meat ducks may be started on 20 to 22 percent protein feed, more than meat chicks are started on. The protein component of chick feed can be increased by adding a high-protein source, such as fish meal. They will soon go for tasty earthworms.

High-protein feeds fuel early maturity and rapid growth, creating the high rate of conversion of feed to meat desired in meat production birds for the highest market return. If continued after the first month, however, they can cause disorders like angel wing and leg malformations, such as those seen in commercial broiler chickens. Angel wing is also called slipped wing, crooked wing, airplane

A child's wading pool makes a good home for day-old Khaki Campbell and Pekin ducklings. Ducklings travel well in the first day after hatching, because they are sustained by the remainder of yolk, and do not require food or water. Ducklings can start on chick starter, but avoid the medicated kind, as they can eat more than chicks and ingest an overdose. *Art Lindgren*

Harvey Ussery's mixed flock enjoys comfrey, both fresh and dried, mixed with feed. Ussery recommends the plant as feed for its mineral and protein content. As a crop, he finds comfrey a tough, resilient plant that ducks and geese are happy to graze themselves. *Harvey Ussery*

wing, and drooped wing; it's a malformation of the last joint of the wing, twisting it so that the feathers stick out. Traditional duck keepers like Cyril Menges favored slower growth with a lower protein diet, not more than 16 percent. Grain like rolled oats or oatmeal in the early weeks can supplement the diet of ducks foraging on pasture. Ducklings like chopped hardboiled eggs, which provide a good source of protein before they are catching enough of their own earthworms, slugs, and insects. At the age of a month or six weeks old, cracked corn, wheat, and milo can be added to the ducklings' diet.

Meat production birds can transition to 16 to 18 percent protein feed at three to four weeks and to 16 percent protein at twelve to thirteen weeks. By then, they can be out on pasture and in the pond. Ducks will get some of their nutrition from the vegetation in a pond, but they will need supplemental feed. A domestic flock probably will not get more than 10 percent of its diet from pasture, and any outdoor location will vary from season to season as to what food it offers ducks. Providing whole oats in addition to pellets and pasture will ensure that your ducks get a balanced and complete diet.

Ducks are omnivores that relish fruit, vegetables, and greens. They will be happy to clean up under your fruit trees. Keep in mind, though, that free-range ducks are likely to destroy garden plants unless the plants are mature and large. They cannot eat mature grass, so you may have to mow occasionally to encourage new growth for the ducks to eat. If you can collect pests like Japanese beetles, the ducks will eat them up. Shake the beetles into a bucket with a couple of inches of water in the bottom and dump the whole thing in the ducks' pen.

Grit, the crushed granite that chickens eat, is important to ducks, too. Duck gizzards enclose it to grind up their food, allowing them to extract full nutrient value from all they eat. Ducks get their own grit if they are on range foraging, but ducks that are confined should have grit available.

During the spring egg-laying season, ducks need calcium. They will help themselves from a dish of oyster shell. Return to regular feed after the laying is done, though, because excess calcium in the diet can cause kidney damage.

These Khaki Campbell hens help Cheryl Lindgren garden by eating up earthworms that come to light. Earthworms provide protein in the diet of ducks that are free to forage on range. Ducks can control pests but are best kept out of the garden until the plants are mature enough not to be part of the ducks' menu. *Arthur Lindgren*

HEALTH MANAGEMENT

Good nutrition and clean living conditions are the best protection against disease. Dave Holderread's books on ducks and helpful sites like Cornell University's Duck Research Laboratory give detailed advice on treating common duck diseases yourself. Develop a good relationship with a veterinarian who is interested in working with ducks so you will have someone to call on when needed.

Keep a first-aid kit to treat common problems. Include first-aid ointment for minor abrasions; medicated eye ointment for eye injuries; and a broad-spectrum antibiotic, such as amoxicillin or tetracycline, for systemic treatment of infections, available at the feed store.

Vaccines are available for the common duck diseases *Riemerella anatipestifer*, duck virus hepatitis, and duck plague (duck virus enteritis). They are generally used only in the event of an outbreak.

PARASITES

Ducks have good natural resistance to external parasites like lice and mites. However, stress, especially from overcrowding, can make then susceptible to infestations. Dirty living conditions set birds up for all sorts of infestations and diseases.

Scratching the head and neck with their feet may indicate lice. Other symptoms include slow growth, weight loss, poor laying, abandoned nests, and irritability. If you suspect parasites, check the bird by parting the head, neck, and vent area feathers in a strong light and looking for the critters. Mites may be on the birds only at night, so you may not see them.

Pyrethrin-based insecticides may be used to control infestations. Homeopathic remedies like olive oil, chewing tobacco, and diluted organic apple cider, rubbed into the feathers, have also been used successfully. Avoid contaminating water and feed. Buildings, nests, and roosts should also be treated.

Internal parasites like worms can be diagnosed by veterinary testing of a fecal sample. If your ducks have acquired worms, they can be treated with commercial wormers. Use chemical wormers only when a problem has been diagnosed; ducks do not need regular doses for preventive treatment.

INJURIES

Treat open wounds by clipping feathers away from the edges of the wound. Wash with clean water and apply antibiotic ointment twice daily. Dust or spray insecticide

Ducks require protection from predators. Their natural feathering protects them from cold temperatures. Arthur Lindgren puts the finishing touches on his duck house in Maine. He and his family keep Khaki Campbells and Pekins for their own use as meat birds. They like the Pekins best and found them ready for slaughter at eight weeks of age. "They were cheap to raise for excellent meat," he says.
Art Lindgren

around wounds to prevent maggots. If the wound is large or there are multiple wounds, give systemic antibiotics per directions.

Large wounds can be stitched, but this may be the time to call in the vet.

HOUSING

Ducks are outdoor birds that do well with a simple shelter, although they do require protection from predators. Bantams are especially vulnerable to predation by owls, so they need a place to hide at night.

Ducks are naturally very hardy in the cold. Any place where they can get out of the worst of the weather is adequate, such as a straw-bale shelter. Remember, their downy feathers have strong insulating qualities. Ducks need shade to protect them from the sun. If you have trees that provide shade, that will be adequate. Alternatively, a canvas or plastic tarp can be stretched between poles to create shade.

An enclosure that keeps ducks dry and allows them access to food and fresh water is adequate. The structure does not need to be more than 3 feet high. Allow 2 to 6 square feet inside per bird. Three solid sides, set against prevailing winds, are enough in most climates. The fourth side can be closed with secure poultry fencing, which ensures adequate ventilation. If the fourth side is solid, make sure enough secure windows or slats allow for air circulation. No additional heat is required. Provide plenty of dry litter and replace it every day or two as needed.

The yard outside should allow 10 to 50 square feet per bird. The ground should be prepared with gravel and sand to drain droppings. Duck droppings are over 90 percent water. The top 2 or 3 inches of sand should be removed periodically and replaced with clean sand as needed. Ducks can become smelly if not kept clean. Locate the water source outside the shelter to keep moisture outside.

Harvey Ussery of Virginia recommends this drown-proof waterer for waterfowl. He places the lower third of a five-gallon plastic bucket, with holes drilled to allow it to settle, upside down in a six-gallon rubber watering tub. A hose attached to a float valve supplies water as needed. "The waterfowl are able to dunk their heads; and chicks who fall in are able to scramble up onto the rim or the upturned bucket bottom, rather than floundering around helplessly in an open waterer," he says. Harvey Ussery

Deep Litter

Ducks can be kept on deep litter, a system that requires removal of the soiled litter only once a year. Starting with a dry dirt floor, cover with 6 to 12 inches of pine shavings, oak litter, or other clean, organic material. Allow 4 to 5 square feet of floor space per bird. As the litter decomposes, add clean litter to keep the depth at about a foot.

Ducks don't scratch up the litter and turn it automatically the way chickens do. You may need to turn the litter with a pitchfork every week or two, to break up the crusted manure on top. If you keep chickens, you can scatter some scratch feed around and turn them out into the duck house while the ducks are outside in the cold. They will be happy to do the turning for you.

Deep litter produces heat, like compost, as it decomposes. Some small-flock keepers report that their ducks lay more eggs when kept on a deep-litter system during cold winters.

Duck feet are subject to injury on metal or wire. Stones under their feet must be smooth. Vinyl-coated 1-inch wire mesh can be used for adult ducks.

Nest boxes constructed with hinged lids allow easy access for collecting eggs. Supplemental lighting can extend the laying season. Use automatic timers to turn on lights before sunrise and after sunset to provide light stimulation for a constant fourteen to seventeen hours a day.

BREEDS

Although all domestic ducks except the Muscovy are descended from Mallards, they are as varied as dog breeds. Wood ducks and Mandarin ducks are small wild ducks related to the Muscovy and recognized by the American Bantam Association. Domestic ducks range in size from bantams that weigh less than 2 pounds to large, imposing birds of 12 pounds or even more. Consider the space you have available and the climate in choosing a breed. Focus on your goals: egg and meat production, preserving a historic breed, exhibition, or personal satisfaction. Visit poultry shows and join poultry organizations to learn about the range of colors and characteristics to find one or more breeds that suit you and your situation.

For best results, choose a beautiful one you love.

Bantam breeds include Call ducks, East Indies, and Mallards. Australian Spotted ducks are increasingly popular but not yet recognized by the APA, so they can be exhibited, but they cannot win top awards. Bantams are good seasonal egg layers and can be desirable small but tasty table birds. They manage well on pasture and effectively control pests by consuming mosquito larvae, snails, slugs, and other pesty insects.

BANTAM BREEDS

CALL DUCKS

This tiny breed began as a decoy, to attract wild ducks so hunters hidden in blinds could shoot them. Their bright, intense voices hark back to that beginning. They are mainly raised for exhibition and as pets. If you raise

These petite Call ducks, a drake in front of three hens, paddle happily on a pond in Texas. Call ducks are recognized in Blue, Buff, Gray, Pastel, Snowy, and White, but fanciers who love these bantam ducks raise them in many other colors. They are often the most plentiful ducks on exhibit. *Kristine Tanzillo*

Odor-free Duck House

Arthur Lindgren of Maine planned to eliminate odors, at least to humans, in building his duck house. By making the floor ½-inch mesh, most droppings fall through. Remaining droppings can be washed through with a hose after the ducks are allowed out in the morning.

Elevating the structure on legs gives the ducks shade and protection from birds of prey. Below the mesh floor, corrugated plastic sheeting in a light frame built of one-by-four lumber catches droppings and wastewater. The sheeting comes in 2½-foot-wide panels. They can slide onto the frame, which is supported inside the legs that elevate the house. Allow panel edges to overlap. Panels can be removed for more thorough cleaning as needed. The area under the house remains clean, except for the droppings of ducks who take advantage of the shade.

Make sure the corrugated plastic slopes enough to allow the waste water and droppings to flow down. Channel the runoff into a length of regular roofing gutter. Drain the sloping gutter into a five-gallon plastic bucket.

The diluted manure can be added to a compost pile or applied directly to shrubs, garden, and fields. Rotating the application avoids burning plants with excessive uncomposted manure.

"The design helps keep the duck-waste odor at near zero to humans," says Lindgren. "But we also have a fairly large duck pen area. We have the space for it so we made it big. The ducks wander all over this area, which allows the droppings to dry and get absorbed into the ground. They don't concentrate their waste droppings in one spot."

Arthur Lindgren

Heavy (8–10 pounds)
  Pekin
  Aylesbury
  Rouen
  Muscovy: Black, White, Blue, Chocolate
  Appleyard
  Saxony

Medium (5–8 pounds)
  Cayuga
  Crested: White, Black
  Blue Swedish
  Buff

Light (3.5–5 pounds)
  Runner: Fawn and White, White, Penciled, Black,
    Buff, Cumberland Blue, Chocolate, Gray
  Khaki Campbell
  Magpie: Black and White, Blue and White
  Welsh Harlequin

Bantam (18–40 ounces)
  Call: Blue, Buff, Gray, Pastel, Snowy, White
  East Indies
  Mallard: Gray, Snowy

your Calls talking to them, they will happily converse with you. On the other hand, not every listener appreciates their quacking. Consider your proximity to the neighbors and the peace of other family members.

Breeders and exhibitors now raise them in many colors. Gray, the original Mallard color pattern, and White were included in the first *Standard of Excellence* in 1874. Buff and Pastel were recognized in 1996. Other colors include White-Bibbed Blue and Black & White Magpie. Chocolate, Pied, Silver Appleyard, and Self Black are working toward recognition. New colors are being developed constantly, such as Butterscotch and Cinnamon.

BLACK EAST INDIES

Slightly larger than Calls, Black East Indies ducks were also included in the first *Standard of Excellence*. Despite the name, some authorities assert that the breed originated in nineteenth-century America. It was often called the Buenos Aires Duck in the nineteenth and early twentieth centuries. Similar ducks have a long history in South America.

Their lustrous black feathers have a stunning green sheen. They are popular exhibition ducks and less inclined to quacking than Calls. Their bodies are longer than the plump, rounded body of the Call.

MALLARDS

Mallards are familiar wild residents and visitors to North American ponds. They are the original domesticated breed, but they were not added to the *Standard* until 1961. The wild greenhead plumage is called Gray. The *Standard* sets a weight of 40 ounces for old drakes, but domestic and exhibition birds often grow larger than wild ones. The Snowy variety, sometimes called the Aleutian, has a long history but was only recognized in 1987. Fanciers also raise White and many other color varieties.

AUSTRALIAN SPOTTED

This attractive small duck was developed from Australian Mallard stock imported to the United States in the 1920s. Young John Kriner, a legendary poultry breeder, worked with the Australian Spotted. He crossed them with birds of several Mallard color variations and a Northern Pintail hen to produce several color variations of the spotting pattern. Henry Miller of Blue Stream Farm developed his own strain in the 1940s from an accidental cross of White Calls and wild Mallards. The breed is not yet recognized by the APA.

In Australia, the Mallard is as popular a small show duck as Calls are in North America. Many varieties have been developed, including the Spotted pattern, which was imported to North America from Australia.

The Australian Spotted duck's weight ranges from 30 to 38 ounces, making it a bantam duck. The variety name refers to the head color of the male. The original variety was seal brown, but Mallard Greenheads, Blueheads, and Silverheads have been developed.

The Australian Spotted duck retains the hardiness of its wild ancestry, develops quickly, and is a good forager. Hens may start to lay as young as three months old and are good setters and mothers.

MANDARIN DUCKS AND WOOD DUCKS

Mandarin ducks are a small, wild Asian breed that thrives in small flocks and tames easily. The breed's numbers are much reduced in Russia, China, and Japan due to habitat destruction and hunting. Feral colonies have been successful in Europe and England. Its colorful plumage has made it an attractive feature of decorative landscaping in American gardens. The Mandarin and the North American wild Wood duck are related, but the two species do not interbreed successfully.

Wood ducks are hunted in North America, and fortunately, their numbers are not threatened. Their population declined sharply in the nineteenth century due to overhunting but has recovered. They are second only to Mallards in numbers taken. They regularly raise two broods a year.

The quiet whistling vocalizations of Wood ducks help make them a more suitable fit in suburban locations than noisy Call ducks. If raised by hand from ducklings, they make personable companion ducks.

Mandarins and Wood ducks are fully capable of flying and should have flight cages and nest boxes off the ground to accommodate them. Otherwise, their wings can be pinioned to keep them at home.

Both are bred in a variety of colors, such as White, Apricot, and Silver, as well as the original wild colors.

LIGHT BREEDS

RUNNERS

The Runner duck is a historic breed that developed in Southeast Asia to run long distances to the rice paddies from its home. Its unusual upright carriage typically stands at 45 to 75 degrees from the horizontal, standing almost straight up when alert. The Runner duck has been described as a wine bottle with legs.

In the wild, Wood ducks nest in natural cavities such as hollow trees. Provide them with secure, secluded nest boxes. In the wild, they may nest very high. Newly hatched ducklings must jump down to join their mother and get into the water. Safe jumps as far as 290 feet have been documented.
Metzer Farms

Runner ducks are popular for their unusual posture and conformation. They hold their long, slender bodies at a 45 to 75 degree angle from the horizontal. They don't waddle like other ducks but instead run smoothly. Their history traces back to the East Indies as far as two thousand years, judging from stone carvings in Java. They are the Leghorns of the duck world, the best egg-laying breed. A hen may lay as many as three hundred eggs a year. *Metzer Farms*

They are popular exhibition birds, recognized in eight colors and bred in many more, from Mallard-type to solid, penciled, and splash. A rare crested variety is called the Bali or Balinese Crested, but it is not yet recognized by the APA.

They lay well, as many as three hundred eggs a year.

CAMPBELL

This duck is named for the woman who developed it, Mrs. Adele Campbell of Liley, Gloucestershire, England, in the late nineteenth century. Her tale, recounted in "Ducks: Show and Utility" by C. A. House and quoted in Holderread's *Guide to Raising Ducks*, tells that she had a single Runner duck. Although it was an unattractive little hen, the duck had distinguished herself by laying 195 eggs in 197 days. So Mrs. Campbell bred her to a larger Rouen drake, and the rest is history.

Later, other breeds were crossed into the line. The aim was to create a buff-colored duck, but breeding Penciled Runners into the mix resulted in a color pattern that reminded her of the khaki uniforms British soldiers fighting the Boer War in South Africa wore at that time. The Khaki variety became the dominant variety and remains the only one recognized by the APA. Dark and White varieties are also well-established in North America.

Campbells retain the active nature of their Runner heritage with a Mallard-type carriage, 20 to 40 degrees above the horizontal. They are resilient and adaptable to all climates.

They and their descendants, Welsh Harlequins, are the best egg layers of all ducks, laying as many as 340 pearl-white eggs a year. Some strains have declined in laying ability, so be sure to acquire birds from reliable breeders. Avoid birds that have stripes on their faces or weigh more than 6 pounds, giveaway indications that other breeds are influencing the bird.

MAGPIE

This duck was developed in the twentieth century from Runner ducks and other influences. Its black-and-white plumage and head, bill, body, and carriage are suggestive of the old Dutch breed, the Huttegem, which may have been involved in its development.

Magpies lay well, as many as 290 eggs a year, and the meat is gourmet quality. They are hardy and good foragers. Despite their fine qualities, the breed remains rare.

These black and white Magpie ducks don't mind the snow at their home at Yellow House Farm in New Hampshire. They take their name from the Magpie color pattern: white plumage with contrasting markings on the head, back mantle, shoulders, back, and tail. The white areas tend to increase with each annual molt. Magpies are a challenging breed to exhibit because of the difficulty of reproducing the pattern, but they make a good general-purpose duck for the farm.
Robert Gibson

Magpies are recognized in Black & White and Blue & White varieties. The color pattern—white with a black cap, back, and tail—doesn't breed perfectly, so getting show-quality birds is a challenge for the exhibition-minded.

WELSH HARLEQUIN

Welsh Harlequin ducks are gaining a following in the United States. They were developed by Leslie Bonner, a Welsh commercial duck breeder, from Khaki Campbells in 1949.

Welsh Harlequin eggs were imported to the United States in 1968. Additional adults were introduced to the resulting two confined flocks in 1981, and ducks have been available to breeders since 1984.

These excellent egg layers produce as many as 330 pearl-white eggs a year and also make tasty meat birds. Their light undercolor feathers are less noticeable in the skin, giving the plucked carcass a clean appearance.

Dave Holderread of Holderread Waterfowl Farm and Preservation Center in Oregon claims to be able to sex the ducklings with 90 percent success by the color of the bill during the first day after hatching. The males have dark bills and the females light bills with a dark tip.

MEDIUM BREEDS

ANCONA

Anconas, with their black-and-white broken-pattern plumage, belie their relationship to Magpies and the old Dutch Huttegem. They are a twentieth-century duck. Like pinto horses, no two have identical patterns of color.

They are good foragers and adapt to all climates. Rich Serfass of Michigan's Upper Peninsula reports that his Anconas thrive in the cold winters there: "I still have one original hen Ancona who raises ten to thirteen ducklings every year and is now going on six years old."

Anconas are the best egg-layers of the medium breeds, laying as many as 280 eggs a year.

Although not yet recognized, Anconas are being raised in at least six varieties: Black, Blue, Chocolate, Silver, Lavender (all with White), and Tricolor (White with any two other colors).

CAYUGA

These glossy black ducks glimmer green in light. Those beautiful feathers can make it difficult to pluck them clean, however. Skinning can avoid the problem of dark pinfeathers.

Cayuga ducks are a good choice for both eggs and meat. They are easygoing and adaptable. Like other waterfowl, they don't mind the snow. They were one of the breeds included in the first APA *Standard* in 1874. Limit sun for show birds, as it can change the green highlights to purple. For showing, the blacker and greener their feathers, the better.
Robert Gibson

These ducks are a nineteenth-century development, named for Lake Cayuga in New York, which was named for the Cayuga people of the Iroquois nation. Lake Cayuga was reputed to be the origin of large black ducks in general. This breed probably resulted from interbreeding between white farm ducks and the American Black duck, a wild breed also known by several other names, including Dusky duck and Black Mallard, which is neither a true Mallard nor black but is actually a dark brown. Cayuga males should have the curled tail feathers typical of Mallard drakes, but American Black duck heritage results in some with straight tail feathers. That's a fault that costs points at showing but is not a disqualification.

Cayugas were included in the first *Standard of Excellence* in 1874. They lay a respectable 100 to 150 eggs a year and produce a meaty carcass. The eggs start out the season almost black, fading over the course of the season to gray, to blue, and eventually to white. They are also popular exhibition birds and pets. A Blue variety is also raised.

CRESTED

Crests appear as a dominant mutation in all breeds of Mallard descent, one in every one hundred thousand to one million eggs hatched. Selective breeding has brought this characteristic into a line that reliably breeds some crested offspring.

The mutation is far from perfect in its breeding, however. Eggs with two of the dominant genes for the crest do not survive. Those with the single gene are subject to other skeletal abnormalities, including kinked necks, shortened bodies, roached backs, and wry tails and may have balance problems. Any given breeding will produce about one-third uncrested offspring. Crests in those that have them may be small or misshapen, regardless of what their parents' crests were like.

The birds are so impressive in appearance, though, that they are pursued and desired. They are usually good egg-layers and make a good roasting bird, but they are mainly raised for exhibition or as companion animals. The APA recognizes Black and White varieties; fanciers also raise Buff and Gray varieties.

BUFF OR ORPINGTON

Although recognized in the *Standard* as Buff, this breed is also known by the English name, Orpington. No other colors are recognized or currently bred in the United States.

Of substantial size and often laying more than two hundred eggs a year, Buff ducks are a good all-purpose breed. They are similar in type to the Cayuga. Their light color makes them easier to pluck for a clean carcass.

The buff feathers tend to fade, so Buff ducks that are intended for shows should be protected from the sun six to eight weeks prior to the show season.

BLUE SWEDISH

This stocky duck originated in the Pomeranian region of Germany, which was part of the Swedish Kingdom in the early nineteenth century. They came to the United States late in that century and were admitted to the *Standard* in 1904.

They are respectable seasonal egg-layers, laying more than one hundred eggs a year. The blue color pattern does not breed true, making refinement of the color for exhibition a challenge. The additional requirement that the outer two or three wing primaries be white adds another level to the exhibitor's goal. Blue feathers are also subject to fading, so birds intended for show should be protected from direct sunlight before and during show season.

As in chickens, the blue color passes to only about half the offspring of blue-to-blue color matings. The others will have either black or silver (very pale blue) plumage.

HOOKBILL DUCKS

This is one of the oldest duck breeds, documented back to the sixteenth century. It's easy to recognize the ancestors of today's birds in Dutch art by their distinctive bills. Today they are rare, but dedicated breeders like Patrick Sheehy of Serenity Farm in New Hampshire are preserving the breed.

In Sheehy's experience, Hookbills do not fly. His birds take flight only when harassed and even then are not capable of sustained flight. He keeps them in open-topped pens.

The hens lay as many as one hundred large blue eggs annually. They weigh 5 to 6 pounds, making them a desirable table bird.

In the United States, Hookbills are raised in three color varieties: Dusky, White-bibbed Dusky, and White. The latter variety has a pink bill.

Society for the Preservation of Poultry Antiquities Duck Critical List

Ancona	Rare	Magpie	Old and Rare
Aylesbury	Old and Rare	Mallard	Old
Australian Spot	Rare	Muscovy	Old
Balinese Crested	Old and Rare	Pekin	Old
Bantam Silkie	Rare	Rouen	Old and Rare
Buff	Rare	Runner	Old
Call	Old	Saxony	Old and Rare
Cayuga	Old and Rare	Shetland	Old and Rare
Crested	Old and Rare	Appleyard	Rare
East Indies	Old and Rare	Swedish	Old and Rare
Hookbill	Old and Rare	Welsh Harlequin	Rare

HEAVY BREEDS

SILVER APPLEYARD

The Appleyard, which exists in only the Silver color pattern, was developed in the 1930s in England to be big, to be beautiful, to lay lots of white eggs, and to have a large, meaty breast. It succeeds on all counts. Reginald Appleyard of Priory Waterfowl Farm in West Suffolk set out to create the ideal duck. He can be proud of the breed that honors his name.

The Appleyard was brought to the United States in the late 1960s but was not available to the public until 1984. The breed has been recognized for exhibition.

In England, both crested and plain-headed varieties are shown, but in the United States only plain heads are shown. Crested ducklings are not uncommon, though.

This breed is adaptable, sturdy, and self-reliant. The hens lay as many as 270 eggs a year and make good brooders and mothers.

Miniature Silver Appleyard ducks are now available in the United States from Holderread Waterfowl Farm and Preservation Center in Corvallis, Oregon.

AYLESBURY

The Aylesbury is the whitest of all ducks, in plumage and skin. In the early nineteenth century in England, where it was originally known as the White English, breeders thought that the local white pumice the ducks consumed as grit was the crucial factor to its color, but genetics and other factors turned out to be more important.

The satin-white plumage of Aylesbury ducks depends not only on their genes but their diet and living conditions as well. Hard water for bathing, color from bedding, and reddish soil with elevated iron levels can discolor feathers. Feed that has yellow and orange pigment, such as fresh greens and corn, will show up in the feathers and skin.

Aylesbury ducks start out with the ideal pink bills, and the bills of laying hens are pink. The bills of non-laying hens and drakes may turn pale yellow with a pink tinge.

Because of their white skin, Aylesburys may need vitamin A and D3 supplements, especially if they are on a restricted diet to keep them white for exhibition. Dave Holderread recommends four parts game-bird flight conditioner, two parts oats, two parts white wheat, and two parts fish-based cat kibble, supplemented with a multivitamin. These additives together should supply the diet with the appropriate amounts of vitamin D3 and vitamin A.

This historic breed was included in the first *Standard of Excellence* and continues to show well.

These Aylesbury ducks winter well on a Kansas farm. Ducks are insulated with down feathers that keep them warm in the coldest weather. The feathers are an important product used in making insulated clothes, pillows, and bedding. *Courtesy of author*

Muscovy hens are excellent egg layers and mothers. Wild birds are black and white, but selective breeding in domestication has produced solid white, solid black, chocolate, and blue varieties. Many find them charming companions. *Shutterstock*

MUSCOVY

Muscovies are native to Central and South America and Mexico, where they were first domesticated. They remain wild in their native habitat. Several different strains, some having more wild characteristics, are available, so make sure you know what you are acquiring. In the wild, they roost in trees.

Muscovies take to domestication well, as they have since before contact with Europe. Columbus may have introduced them to Europe along with the turkey, or Spanish explorers may have brought them from South America to Africa in the sixteenth century; from there, they would have traveled along trade routes to Europe. They were exhibited at the first American poultry show in 1849, and they were included in the first APA *Standard of Excellence* in 1874.

Muscovies are excellent natural layers, the hens often laying as many as twenty eggs in a clutch and raising two clutches in a season. They are good mothers and are often used to incubate eggs of other birds.

Many small-flock meat producers prefer Muscovies for the table. They grow quickly to market size, but overfeeding them to increase growth can cause leg and reproduction problems.

The warty growths on the head, called caruncles, are unique to Muscovies and an important show point. For

Muscovies are different from other domestic ducks, which are related to Mallards. Muscovies have a long history of domestication in Mexico and Central and South America before European contact. The caruncles are typical of the breed. Males have more than females. *Shutterstock*

exhibition, the caruncles should be equally distributed on both sides of the head, not so extreme that they interfere with the crest feathers on top of the head or the bird's vision.

Muscovies are personable and self-reliant, although some are inclined toward aggression. Many small-flock owners warm to their sociable personalities. Females may be inclined to fly, although males are typically too heavy. Clipping primaries may be necessary, although they are faithful to a good home. They are the quietest duck. Muscovies retain the strengths of their wild past in domestic life.

PEKIN

Pekins got their name from the Chinese city Beijing, formerly known as Peking. The breed's history of domestication extends back more than one thousand years. They remain the most widely raised commercially.

Pekins came to the United States in 1873, in time to be included in the first *Standard of Excellence*. They have the fastest growth rate, reaching as much as 8.5 pounds in seven weeks in a strong body. The hens lay well, as many as 225 eggs annually, with high hatchability rates. Little wonder their numbers exceed all other ducks combined.

Unlike with the Aylesbury, yellow in its white plumage is desirable, so greens and corn are welcome in the Pekin's diet, even for show birds.

They are sociable and make good pets as well as attractive birds on the pond. The hens are especially talkative.

ROUENS

This breed began in France several hundred years ago, where it acquired its name. Like the Toulouse goose, it was developed for the culinary advantages of its large liver and its desirability as a roasting bird. The name has become synonymous with large, gray table ducks, whether or not they have any actual breeding connection to French ducks. "Commercial Rouens" may not have any relationship to true Rouens. Development in England and the United States has created several different strains. Production strains are smaller, topping out at 8 pounds. Exhibition strains grow as large as 12 pounds, although the *Standard* top weight is 10.

Both production and exhibition strains make good choices for small flocks. The large size of exhibition birds may make mating and setting difficult. Six inches of swimming water makes mating more likely, but large males may not be very fertile. Large hens may break eggs, so these may need to be incubated artificially or by another hen.

The Rouen has Mallard-type plumage and undergoes an eclipse molt like the Mallard.

Eclipse Molt

All ducks molt in June, after the breeding season ends. Male ducks of many breeds in the Mallard line have much brighter and more colorful plumage than the camouflage plumage of the females. That's typical of ground-nesting birds, since it protects the females on the nest. Once the breeding season is over, however, the males lose those bright feathers and take on the plumage of the females. This is called an eclipse molt.

Mallards then molt again during the months of September to November, when the males acquire their winter plumage, which takes them through to the next breeding season.

When eclipse molt occurs, the birds are flightless, since they lose all their flight feathers at once. The drab plumage gives them camouflage while they are vulnerable. Wood and Mandarin ducks also experience an eclipse molt.

Ducks socialize compatibly with other waterfowl and poultry in mixed flocks. Harvey Ussery's African geese and Saxony ducks flock together on the Modern Homestead in Virginia. Taking advantage of their natural behaviors to maintain his farm, he keeps mixed flocks of waterfowl to control insects and slugs and clean up orchard waste. *Harvey Ussery*

SAXONY

Albert Franz of Chemnitz, Czechoslovakia, is credited with developing the Saxony prior to World War II. He bred three breeds—the Rouen, the German strain of Pekin, and the Blue Pomeranian—to create an attractive, sturdy, productive duck.

The Saxony nearly disappeared during the chaos of war in Europe, when most were consumed as food by desperate survivors. Herr Franz recovered his breeding program after the war, and the breed was recognized in Germany in 1957.

The Saxony's tight feathers cover a muscular body, and it is known as an excellent roaster. The rich colors of its plumage may fade in direct sunlight, so keep them in shade if you intend to show them. The drake's Mallard plumage is accented with a collar of white around the

Egg Duck Breeds

Breed	Eggs/year
Heavy	
Pekin	225
Aylesbury	125
Rouen	100–150
Muscovy	100
Silver Appleyard	270
Saxony	240
Medium	
Ancona	280
Cayuga	100–150
Blue Swedish	100
Buff	200
Hookbill	100
Light	
Runner	300
Khaki Campbell	340

Egg Breeds

Many duck breeds lay as well as chickens, but generally, ducks are seasonal layers. Like chickens, their laying is influenced by increasing or decreasing daylight. Adding artificial light can encourage them to lay longer. Ducks also require adequate nutrition, 16 to 20 percent protein, to be good layers. Strongly flavored foods, such as fish meal, will influence egg flavor.

When cracking a duck egg, the membrane lining the shell may be tough. Nick it with the point of a knife after cracking the shell, and it will tear.

Ducks will naturally start laying in the spring and lay through the summer, tapering off in fall until they stop laying in the short days of winter. Good egg breeds are Runners, Harlequins, and Campbells, all light breeds. Ancona ducks are a good medium egg breed and Appleyards, Pekins, and Saxonys are good heavy breeds. Muscovy ducks, also a heavy breed, are good natural layers, often laying twenty or more eggs in a clutch, compared to the twelve to fifteen typical of Mallard-type ducks. They are also willing to lay additional clutches.

I grew up eating Muscovy eggs, so I know they are delicious.

neck. The hens are creamy buff with white throats and white stripes on their faces.

Saxony hens may lay as many as 240 eggs a year, the most among the heavy breeds. They are strong, resilient birds that are good foragers.

EGG PRODUCTION

Ducks can be good egg layers with some supportive nutrition and husbandry. Ducks on pasture without attention to management for egg production will not reach the numbers listed in the table [*page 52*]. For maximum results, choose birds carefully and care for them with egg production in mind.

Like most poultry, ducks are naturally seasonal layers. They lay in the spring, as do wild ducks. Muscovy ducks differ from Mallard-type ducks in naturally laying and raising multiple clutches during the course of the year.

However, if eggs are collected and the hen is not allowed to accumulate a clutch, she will continue to lay. That's the heart of egg production. Breeders who then select the hens that continue to lay the most eggs and breed them will have increased egg production in their flock.

Most breeds have separate strains that, while they meet the same physical breed characteristics, differ in significant ways. One of those is egg production. Ask whether the birds you are considering acquiring are an egg-producing strain.

Egg production is physiologically demanding on a duck. She must have high-protein feed, about 19 percent, with additional calcium to support her. She can't lay what she doesn't have (i.e., calcium for the eggshells, and protein and water for the contents). Overweight ducks produce fewer eggs, so monitor weights to avoid fat ducks.

The physiological trigger that starts egg-laying in the spring is the lengthening hours of daylight. Photostimulation is necessary for ducks to continue laying eggs. Increase and maintain day length by half an hour a week to a maximum of fourteen to fifteen hours of daylight or artificial light to encourage egg production.

CHAPTER 4

GEESE

Geese are generally hardy and easy to manage. They are naturally resistant to the maladies that afflict other poultry. Reginald Appleyard, legendary English waterfowl breeder, describes them as "being amongst the brainiest of all classes of domesticated fowls." As grazers, they are not competitive with each other for food but instead form a cohesive gaggle. Domestic geese retain some ability to fly, but they need time to take off and a clear runway. With a happy home and comfortable living conditions, they are unlikely to present any problem by taking to the air.

Domestic geese retain some wild qualities. Even wild geese tame relatively easily. Wild/domestic hybrids are not uncommon. Domestic geese, like their wild relatives, are seasonal egg-layers. Chickens and some ducks have been selectively bred and domesticated to be year-round egg layers. Geese have not, although China geese are good natural layers and will produce up to one hundred eggs during the season. Some other breeds lay between twenty and forty eggs in a season.

Some geese are territorial, especially during the breeding season, and will sound the alarm when strangers approach. They vary by breed and individual. Dr. Tom T. Walker of Texas finds his Cotton Patch geese remarkably friendly and easygoing: "I have people express surprise that the Cotton Patch geese are so accepting of strangers."

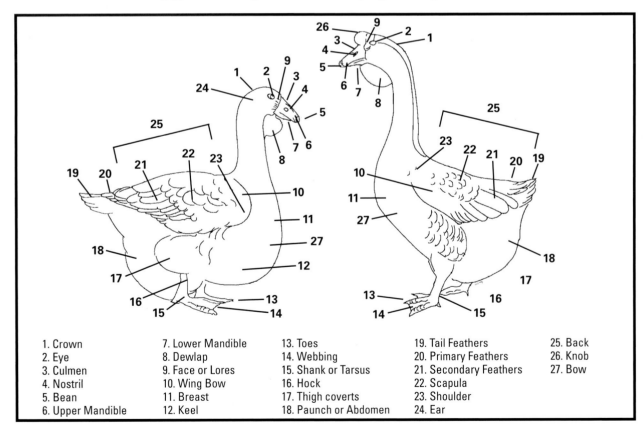

1. Crown	7. Lower Mandible	13. Toes	19. Tail Feathers	25. Back
2. Eye	8. Dewlap	14. Webbing	20. Primary Feathers	26. Knob
3. Culmen	9. Face or Lores	15. Shank or Tarsus	21. Secondary Feathers	27. Bow
4. Nostril	10. Wing Bow	16. Hock	22. Scapula	
5. Bean	11. Breast	17. Thigh coverts	23. Shoulder	
6. Upper Mandible	12. Keel	18. Paunch or Abdomen	24. Ear	

The Nile River in Egypt remains a major flyway for migratory waterfowl, as it has been since pharaonic times. These Egyptian geese are painted on the wall in the tomb of Nefermaat and his wife, Itet, at Meidum. The full panel, nearly six feet long, shows a symmetrical arrangement of these two geese facing left and the other two facing right, bounded by a goose at each end feeding on the grass. They demonstrate the careful observation Egyptians made of wildlife as well as their artistry. The paints were derived from natural materials: white from limestone, red from hematite, green from malachite. Paints were mixed with egg white, which has lasted well for these 4,500-year-old paintings. The panel is on display at the Egyptian Museum in Cairo. *Robert Partridge,* Ancient Egypt *magazine*

Geese make effective watchdogs because they announce the presence of strangers so noisily. They are protective of the flock. Harvey Ussery of the Modern Homestead in Virginia witnessed a gaggle of geese defending the hens and chicks who shared their pasture from a hawk. "They converged, honking in outrage, ready to take on the intruder," he writes. "[The intruder] quickly concluded he was badly out-matched, wheeled in a tight mid-air U-turn, and flew off looking for easier pickings." Lyn Irvine says in her 1961 book *Field with Geese*, "If geese were good for nothing else, I would value them to keep cats humble and spoil their hunting."

On the other hand, neighbors living nearby may object to breeds that are especially noisy, such as Chinese and African Geese. Jeremy Trost of Wisconsin remembers watching his mother in the garden, shouting through the fence at his noisy Chinese geese. The birds, two hens and a gander, were objecting to her presence in the garden, and she was responding in kind. The gander paced along the fence, fuming, listening silently while she shouted at them, then honked back at her. The final confrontation, Jeremy's mother facing off two feet from the gander's face, resulted in the gander tumbling over the fence, to his own surprise.

"He went one way, and my mother went the other," Jeremy remembers.

Jeremy went out the door to capture the gander as his mother escaped to the house. After that episode, Jeremy moved the Chinese to the yard farther removed from the garden. The calmer Pilgrim Geese were moved into the garden yard.

"After all that, I still miss the geese," says Jeremy, who now keeps chickens. "There is something about them."

Many people are devoted to geese, and geese often make pets of people. They even preen their favorite people, a service they do not provide to other geese.

HISTORY AND CULTURE

Geese were domesticated as far back as five thousand years ago in Egypt, the natural flyway for waterfowl migrating between Africa and Eurasia. The migratory flocks included Asia's Swan goose and Europe's Graylag goose, the ancestors of modern domestic geese, as well as the Egyptian goose, technically not a true goose. Egyptians netted them as hundreds of thousands settled on the Nile during their migration. From catching wild birds for food, it's a short step to keeping them in pens, then breeding them and selecting breeding birds for the qualities most desired. Religiously, the goose was associated with the cosmic egg from which all life was hatched. The god Amun sometimes took

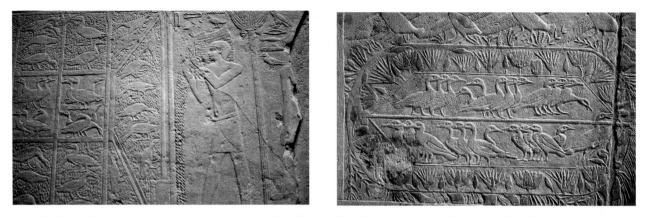

Egyptian bas-reliefs featuring geese decorate the tomb of Kagemni at Saqqara, near the Pyramid of Teti. Kagemni was the Vizier to three kings in the sixth Dynasty circa 2345 to 2181 BC. His tomb is famous for its finely carved scenes. The hunting scene shows wild birds being caught in a clap trap (*left*). The other scene shows an aviary or netted enclosure where wild birds are held and being fed, perhaps being fattened for eating (*right*). Like most scenes in the tomb, they reflect both daily life and also areas over which Kagemni would have had some control as Vizier.
Robert Partridge, Ancient Egypt *magazine*

the appearance of a goose. Geese were also associated with the Egyptian god Geb, sometimes depicted with a goose as his headdress, the green god of the earth, and his daughter Isis, "the egg of the goose" and sister-wife of Osiris, god of the underworld.

The ancient Romans and Greeks raised geese and honored them. Geese were sacred to Juno, queen of the gods, wife of Jupiter and protector of Rome. White geese lived in her temples. They are said to have saved Rome from an attack by the Gauls around 390 BC by raising the alarm and awakening the guards. They became associated with Juno as symbols of marriage, fidelity, and contentment at home. The Greek goddess of love, Aphrodite, was welcomed by the Charities, whose chariot was drawn by geese.

Christian Saint Martin of Tours (fourth century AD) is the patron saint of geese. Goose is traditionally served on his feast day, November 11. The tale is that he did not want to become bishop, so he hid in a barn with the geese. They noisily drew attention to him, and he became bishop of Tours in AD 372. Charlemagne encouraged goose husbandry in his empire, AD 768 to 814.

Celtic myths associated the goose with war, and remains of geese have been found in warriors' graves. Geese also symbolize movement and spiritual quest. Their return each year is a reminder to come home.

Mother Goose may have been based on a historic person or may be a mythic character to embody storytelling. In either case, the goose symbolizes communication, expressing themes of human life in legends and tales. The first book of Mother Goose stories was published in Boston in 1786. "The Goose Girl" was included in *Grimm's Fairy Tales* in 1815.

As recently as a century ago, people in England kept geese in a half-wild state, letting them forage and live on the river. The geese spent the spring and summer on the village green, then migrated to the River Cam for the winter. In February, the owners would call their geese, which responded to their owners' voices and returned home to nest and rear their young. Those offspring were a significant contribution to the villagers' income.

GOOSE IN THE AMERICAN DIET

Geese do not accommodate intensive husbandry and have lost favor in modern America. The reputation that they are difficult to cook and that the meat is fatty have also reduced their appeal. Americans consume much less goose than they did a century ago, when every farm raised some and the goose was the traditional holiday bird. Current USDA statistics show that the average American consumer eats less than a third of a pound of goose annually.

African/Toulouse cross geese enjoy life on a small pond. The floating platform built by their owners gives them a place safe from predators to rest and eat. Although the geese didn't understand the floating platform at first, they did know Evelyn. Her parents put her life vest on her and set her on the platform. "She hollered to them, and they jumped aboard," recalls her mother, Rhonda Andrade. Stocking the platform with food for two weeks encouraged them to visit it. By that time, the geese had adapted to their new home and were fully independent.
Rhonda Edwards Andrade

Contemporary chefs are rediscovering this favored bird on the table. The meat itself is not fatty. As with other waterfowl, the fat resides in a layer under the skin. The fat can become part of the roasting process by pricking the skin to allow fat to drain off during roasting, or the goose can be parboiled before roasting. Goose grease is an underappreciated oil that can be used in baking. Collect it from the roasting pan and use it throughout the year. National Public Radio commentator Bonny Wolf calls it "the crème de la crème of fat."

Geese Prefer Peace

Geese prefer quiet conditions and predictable, calm treatment. Noises and strangers disturb them and can reduce fertility, weight gain, and egg laying. Situate your geese in a protected location where they can settle down and enjoy life. They have little sense of humor and will respond to teasing by becoming aggressive. Don't allow anyone to provoke your geese.

HOUSING

A total pen area of 2,500 square feet should be adequate for a small flock of less than ten geese. If it can include a pond of 500 square feet of water, so much the better. Geese enjoy splashing in water and swimming, although they can manage without it. They stay cleaner and have fewer parasites if they have access to swimming water, and it's easier for the geese to walk to the water than for you to bring the water to them.

The water must be kept clean, despite the geese defecating in it and splashing mud around. Cement-lined artificial ponds or children's plastic pools are easy to clean and don't turn into mud holes, but small wetlands can be constructed and managed to enhance habitat for domestic geese as well as wildlife. Natural running water, such as a stream on your property, can provide the regular fresh water geese need.

Geese tend to be territorial and aggressive in the breeding season, so plan to separate them into individual pens. Like all domestic fowl, geese are vulnerable to predators. Fence them from predators with four-foot poultry wire fencing. In mild climates, security from predators is all the protection they need.

In cold climates, simple structures are adequate to protect geese from the weather. Stacked hay bales with a plywood roof facing south or a semicircular windbreak of straw bales will keep them out of the wind and snow.

Getting Fat

As grazers, geese naturally eat constantly. In confinement, they are inclined to overeat the more concentrated commercial rations and get fat. Avoid allowing them to consume too much food. Being overweight isn't any more desirable for geese than it is for humans. Fat geese may be less fertile. Slim and graceful is the ideal.

Provide plenty of dry litter for them, either wood shavings or straw. Replace it as it gets wet, daily if necessary. As long as geese are well-fed and have clean bedding, their natural insulation can take almost anything winter throws at them. In a winter storm, they may be out looking around while other fowl are sheltered indoors. "I have yet to see a goose get under shelter to get out of the rain!" says experienced breeder Dr. Tom T. Walker of Texas.

A house to lock geese up overnight should provide about 10 square feet of space for each bird. Geese confined for longer periods of time should have 20 square feet per bird. A low shelter open on all sides can offer shade and protect food from moisture.

Domestic geese do not fly much. If flying becomes an issue, trimming four inches off the leading four or five primaries of one wing will prevent them from successfully flying away. Feathers will need to be trimmed again after each molt. Pinioning removes the entire first joint of the wing, cutting it off. It can only be done on goslings in the first day or two after hatching. Pinioning makes it impossible for the bird ever to fly.

FOOD AND NUTRITION

The diet should include grass, since geese are primarily grazers. They enjoy greens from the garden or the local produce department. A friendly produce manager may be willing to save green trim for you. Geese are so good at eating grass that Cotton Patch geese take their name from the job that was theirs on the farm. Sources from the mid-nineteenth century advocate keeping geese primarily for their ability to clean up pasture, with their meat as an added benefit. Ms. Irvine says, "No other creature so rapidly turns grass into flesh—the commonest weed into the most coveted food." F. J. S. Chatterton, in his 1951 book *Ducks and Geese and How to Keep Them*, recommends keeping geese "as a means of improving poor grassland." Geese can be turned out in fields after harvest to glean and clean. They are vegetarians and will look with disdain, as only a dignified goose can, on the relish with which ducks devour insects and snails.

Geese and Snakes

Although geese are grazers and vegetarians, reports of them eliminating undesirable critters from the field are frequent. Dr. Tom T. Walker of Texas says his experience supports reports that geese will banish snakes.

"All my life I have heard that geese will rid a place of poisonous snakes," he says. "When we moved to the place where we now live in 1985 there were many Copperhead snakes. Regularly, they were biting our dogs.

"With the geese, the snakes disappeared. I have not seen a Copperhead snake in eight or ten years. In the process, one goose was bitten on her head. Her head swelled up to about the size of a baseball; but she did not die. Since I never saw dead snakes around, I assume that the geese ate the snakes as peafowl do."

Alfalfa pellets and grain may be added to the diet. Commercial preparations like Purina's Flock Raiser provide complete nutrition. A good breeder ration can be made from equal parts of the following ingredients: high-protein (18 percent) rabbit pellets, 20-percent-protein layer pellets, and wheat. Breeder ration should be started six weeks before hens start laying, around February. Hard foods, such as corn or bread crusts, can be soaked in water. Geese themselves may place hard food objects in water to soften them.

Geese require drinking water deep enough that they can submerge their heads, five or six inches, to clean their eyes and nostrils. Watering troughs set on wire-covered platforms can minimize muddy conditions around the trough. A hard rubber trough 4 feet by 2 feet and 1 foot deep makes a good pool, does not require much water, and is easy to clean.

BREEDING, HATCHING, AND RAISING GOSLINGS

Overall size, length of neck, size of knob, and even voice can indicate gosling sex to the experienced eye. It's a management advantage to be able to tell the sexes apart, for identifying prospective breeding birds. It's disappointing to set a breeding pen only to find out later that they were all of one sex. If you aren't sure, ask for advice. Waterfowl males have intromittent penises that can be identified. Autosexing breeds hatch with distinctive coloring, so they can be identified as soon as the bill—greenish black for a female, brownish yellow for a male—pips through the shell.

Geese, even goslings only a few days old, can be sexed by examining the vent. Males have a penis and females a small genital structure. Techniques for everting the cloaca to examine the genitalia are best learned from an experienced handler. Birds can be permanently injured by harsh handling. Check your local poultry organizations for an expert who is willing to teach you.

Geese are preparing for the breeding season from September to January. They should be in perfect condition by mid-February. They may begin laying from mid-February to mid-March. The neck feathers of the females should be ruffled during the breeding season, indicating that the gander has bred her. If she has smooth, tidy feathers on her neck in March, it is not a good sign.

Geese will mate for life and usually make good parents. "For geese, it's all about family," says Craig Russell, president of the Society for Preservation of Poultry Antiquities. They may take an interest in breeding in their first year but do better if they wait until they are two years old. A single Chinese or African gander may be introduced to

A Blue Steinbacher goose reigns from the nest she and her mate made from straw and hay and lined with down removed from her breast. She is one of the birds imported from Europe by Bernd and Marie-Anne Krebs of Michigan. Mrs. Krebs observes that they are determined brooders that will decline in health if not allowed to set their own eggs. They require a secluded, peaceful location for their nests. *Marie-Anne Krebs*

A Blue Steinbacher gander protects his goose as she sets on the nest. He stretches his strong neck forward between her and the photographer, who reports from experience that "They are like pit bulls. They bite hard and don't let go." She advises not to make eye contact with a defending gander. "Proceed quickly and don't turn your back on the gander," she says. *Marie-Anne Krebs*

as many as six hens and bred to them in succession for an extended breeding season; however, such matings may not succeed. Geese tend to select mates and pair off in the breeding season. Toulouse ganders, for instance, are unlikely to breed more than two hens.

Dr. Walker has observed that the gander may ignore or even reject other mates. "The gander may give all of his attention to one goose, even to the extent of seemingly telling one that she may not swim with him and the other female," he says. In other cases, they all get along well.

Expecting too much from a gander may result in phallus prostration, in which the gander is unable to retract his penis. This can be treated with gentle cleaning and applying medicated ointment until it retracts, but the treatment is not always successful. It may be wiser to eliminate him from breeding, since the weakness may be genetic. This does not happen to birds in mated pairs.

Select strong birds without defects for breeding. Wing problems, such as angel wing, may be caused by environmental factors, but it's wiser to avoid breeding birds that have them. Weak legs are another reason to keep birds out of the breeding pen. Size is less important than type in selecting breeding birds. It is easier to breed for larger size later than to correct defects in type.

Goose Regulations

Wild ducks, geese, and game birds are protected by the Migratory Bird Treaty Act and regulated by the U.S. Fish and Wildlife Service and state agencies. Check with your state to comply with the laws that apply.

For instance, in California, it's illegal to capture any game bird in the wild and confine it. Special licenses are needed for wild bird rehabilitation. A game breeder's license is required to raise domesticated game birds that normally live in the wild in California.

Geese will make their own nests on the ground. Dr. Walker provides a small doghouse for geese in nesting pens but finds they often prefer to nest outside the house. The dampness is important in incubating the eggs. "The goose will even take the hay out of the house and mix it with sticks, leaves, and other things she finds to build a nest outside the nice house that I have built," he says. Hens will line their nests with their own down. Watch them carefully until you are sure the goose will be broody and the pair can manage their nest. Ms. Irvine attributes to the eighteenth-century French scientist Georges-Louis Leclerc, Comte de Buffon, the observation that "the condition of a sitting hen, however insipid it may appear to us, is perhaps not a tedious situation but a state of continual joy." If the goose doesn't cooperate, broody chicken hens or artificial incubators can be used. Many goose eggs are successfully hatched under chickens. A chicken can manage from four to six goose eggs and can foster the goslings. Goose eggs benefit from moisture, as they would receive from their mother on her daily ablutions. Ms. Irvine dunked her surrogate chicken's lower regions in water as she returned to the nest each day.

A typical clutch is ten to fifteen eggs. If the eggs are removed, many geese will continue to lay, as if for a second clutch. For example, Cotton Patch geese typically lay between nine and eleven eggs. A clutch of thirteen or fourteen is exceptional, more than some geese will be able to cover for incubation. If those eggs are removed, leaving an artificial egg may encourage her to continue laying. Others will not lay any more, even if they end up sitting on the false egg alone, trying to hatch it.

Eggs can be stored as long as seven days if they are to be incubated in an artificial incubator, up to four weeks if they are to be incubated by a surrogate chicken. Candle eggs between eight and fourteen days of incubation. Infertile eggs will be clear, whereas developing embryos will show a web of blood vessels early on. The developing embryo will be apparent later. Goose eggs typically hatch within twenty-nine to thirty-one days, but the incubation period may vary from as short as twenty-seven days to as long as thirty-three.

Helping goslings out of the egg is always controversial. Appleyard recommends assisting only if the gosling gets its whole bill out of the egg and then gets stuck. In that case, he suggests pushing the bill back inside the shell and pasting a piece of eggshell over the hole, to give the gosling a chance to make its way out again.

Goslings will start eating grass right away, which can be supplemented with crumble. If a chicken hatches them, she may attempt to feed them as she would chicks, but they will ignore her. Don't feed goslings medicated chick starter. They may consume more than the recommended dose, and it can make them sick.

The floor should be covered with some kind of rough material that gives the goslings' feet something to grip. Otherwise, they may develop leg problems.

HEALTH MANAGEMENT

Geese are sturdy birds not inclined to health problems. Good diet and clean living conditions are the best preventive care. They are long-lived. Ms. Irvine says, "Here in East Anglia, they say that the ninety-ninth year with a goose is the best." That may be an exaggeration, but reports of thirty and forty years are reliable. As with all less-common poultry, develop a relationship with a veterinarian who is interested in geese so that you will have someone to call when you need help.

Sick birds will be weak, lying down, with ruffled feathers and heads down or necks curled around their bodies. Isolate sick birds promptly for treatment to avoid any contagions infecting the flock.

A first-aid kit should include antibiotic ointment and popsicle sticks or tongue depressors to use as splints for broken legs and wings. Set them cautiously and check frequently to make sure you are not constricting the blood supply. Treat abrasions and other wounds with antibiotic ointment, then dust or spray a pyrethrin-based insecticide around the wounds to keep maggots from developing.

Never catch or carry geese by the legs. It can cause permanent damage, even cripple them. Geese are powerful birds and can injure you with their wings, nip you with their bills, or scratch you with their toenails. When you need to catch a bird, herd it into a corner. Catch it by the neck, just below the head. Their necks are strong. If it backs up, press forward and slide your arm under the bird, grasping its legs in your hand. Cradle it on your arm,

with the head pointed behind you. If the bird approaches you in a menacing way and you are determined that you need to catch this bird, lean toward it and grab its wings at the point closest to the body. Good luck.

Always carry a goose with the head directed away from your own face and the back end pointing outward, since the goose is likely to defecate.

A goose that is accustomed to human contact can be handled by surrounding it with your arms and picking it up. Hold the wings close to its body. Keep both goose and yourself safe.

PRODUCTS

The primary product of geese is the table-ready bird for roasting. All geese are good meat birds. Crossing China geese with Embdens makes an excellent meat bird that lays well into September. Table birds are usually butchered before they reach six months of age.

Birds can be processed on the farm or at local processing facilities. Some local governments offer mobile processing facilities built on trailers that can be rented for home use.

To avoid pinfeathers in the carcass, butcher goslings at nine to twelve weeks, before they molt their juvenile

A feather is a flat, two-dimensional object, while down is three-dimensional. Its fluffiness allows it to trap air, preventing transfer of heat and providing its insulating qualities. The filaments interlock inside a quilt or garment. They don't clump together the way synthetic fibers do. Shaking down-filled garments fluffs up the down, adding the air that provides insulation. *Eurasia Feather Inc./Down Inc., courtesy David Sweet*

Live-Plucking

Geese have been live-plucked of feathers and down for household use since colonial times. The practice of live-plucking takes many forms, some more humane than others. The traditional practice involves pulling out some—not all—of the down and small feathers on the breast, belly and neck. Practitioners use thumb and forefinger to pinch a small amount of down and pull against the direction of growth. Large feathers are not removed, since they are not used in making clothing or bedding.

Traditional live-plucking mimics the behavior of setting geese, who naturally remove down from their breasts to line their nests and leave bare a patch of skin. The down lines the nest and allows the warm skin to incubate the eggs. Otherwise, the down would keep the warmth next to the goose's skin and prevent it from reaching the eggs.

Breeder Tom Walker, who grew up in Arkansas during the Great Depression, remembers his mother live-plucking birds in the traditional manner. Geese were plucked of some of their winter feathers in spring and occasionally a second time in August, as the birds were getting ready to molt naturally. "We still use the pillows that Mama made with the feathers that I helped her pluck when I was a child," says Walker.

On the other end of the spectrum, commercial operations in Europe and Asia employ brutal techniques of live-plucking. Pluckers rip out handfuls of feathers, leaving geese bloodied and, occasionally, dead of shock. David Sweet, director of operations for Eurasia Feather Inc./Down Inc., does not purchase live-plucked feathers for this reason.

Relative Desirability of Feathers and Down

Waterfowl feathers and down are different from land-fowl feathers. They developed to insulate the ducks and geese from water. Chickens have an undercoat, but it is not as warm as the down of ducks and geese.

Down is used to fill clothing, comforters, pillows, and mattresses. Ducks are slaughtered for meat at 45 to 120 days old, and their feathers and down are harvested. Geese are usually raised 180 to 800 days, or 6 months to more than 2 years. The longer time to develop allows the birds to grow nicer feathers.

Feathers and down are classified into **quills**, from the breast and tail; **down**, from the undercoat of the breast and belly; and **feathers**, the remaining 60 to 65 percent.

Generally speaking, goose down produces a lighter, more attractive product than duck down. Down can be mixed with steam-curled chicken or turkey feathers. Eiderdown comes from wild Eider ducks. Individual quality depends on the health and nutrition of the birds, their access to clean water and clean living conditions, and treatment of the feathers when they are collected.

This presentation of pâté de foie gras garnished with beet gelee is a gourmet treat made of goose or duck liver. Force-feeding ducks and geese to produce fatty livers is controversial and has been outlawed in some countries as inhumane. Alternative methods of free-feeding take advantage of the birds' inclination to eat enough to gain weight in preparation for migration in the fall. Relying on the birds' own appetites produces similar, although not identical, results. It is available seasonally, in winter only. *Shutterstock*

feathers for adult plumage. Part the feathers and check to see whether pinfeathers are forming. If they are, delay butchering another six to ten weeks until the birds have their full adult plumage. Geese, like ducks, can also be skinned, eliminating the problem of pinfeathers altogether. Poultry wax is a paraffin or beeswax product that can be used to clean the carcass of feathers. Melted, it can be poured over the bird, or the carcass can be immersed in a pot of melted wax. The carcass is then cooled in cold water and the wax pulled away, with the feathers stuck in it. Dark pinfeathers may require a second waxing. Wax can be melted down, strained to remove feathers, and reused. Feathers can be saved after plucking, washed, and used or sold.

Check state laws on selling birds. Every state allows a small number of geese to be sold within the state, but crossing state lines requires USDA-certified processing.

Goose feathers and down are the original insulating materials for warm clothing and bedding.

Goose eggs have the reputation of being superior for baking. However, the white, or albumen, is thicker than that of chicken eggs and may be disappointing for whipping uses. It is not light enough to whip up well.

The gizzard, heart, and liver are all desirable meats. Goose liver is the prime ingredient in pâté de foie gras. Force-feeding geese to develop the fatty livers used in making the delicacy has attracted the attention of animal advocates. The practice remains controversial, and some local bans have been put into place. Fortunately, many gourmands recognize that geese do not need to be forcefully fattened to produce delicious livers.

BREEDS

The three heavy goose breeds recognized by the APA for exhibition—African, Embden, and Toulouse—are impressive, stately birds. They weigh in at 18 to 26 pounds. All have long histories in the United States and were included in the first *Standard of Excellence* in 1874. The recognized medium geese are the American Buff, Pilgrim, Pomeranian, and Sebastopol. Their weights range from 12 to 18 pounds. The two recognized light goose breeds, the Chinese and the Tufted Roman, weigh 10 to 12 pounds. The two recognized ornamental goose breeds, the Canadian and the Egyptian, have been tamed from the wild but are not truly domesticated. They are not classified by weight. In addition, the Egyptian is not a true goose; it's biologically classified as a Shelduck.

Chinese and African geese are descended from the wild Asian Swan goose. American Buff, Pomeranian, Sebastopol, Embden, and Toulouse are descended from the European Graylag goose. All show some influence of the wild Bean goose. Pilgrim geese are a modern composite developed from traditional Gray geese and the old West of England geese.

Many unrecognized goose breeds are attractive and useful. The United Nations Food and Agriculture Organization has identified ninety-six breeds or genetic groups of geese worldwide. The traditional American Gray goose, a larger domesticated version of the Western Graylag, has never been formally recognized but has been the dominant breed raised in the United States since colonial days. It continues to be successful as a market goose for commercial operations. It's often referred to as a "Commercial" or "Production Toulouse"—confusing names, since it is not a Toulouse. It is different in conformation, lacking the Toulouse's fatty keel, and it matures faster.

The American Gray's conformation is the same as that of the American Buff, which was derived from it. Traditionally, it had an orange bill and pink legs, like its wild relative. Orange legs are considered a twentieth-century development resulting from crossing with other breeds.

Recognized Goose Breeds

Heavy (15–26 pounds)
African
Toulouse
Embden

Medium (10–18 pounds)
Sebastopol
Pilgrim
American Buff
Pomeranian

Light (4–10 pounds)
Chinese
Tufted Roman
Canada
Egyptian

Although the Gray goose is plentiful as a commercial bird, flocks that retain the traditional virtues of natural reproduction and good parenting are not. Industrial production makes such qualities irrelevant or even undesirable. This breed could benefit from breeding with attention to the basics of size and conformation. Formal attention could bring it to long-overdue recognition in the *Standard*.

AUTOSEXING BREEDS

Females and males of most goose breeds are similar to each other, but in autosexing breeds the sexes have different plumage. This is known as sexual dimorphism. Ganders are white, while females are solid color or saddlebacked. Saddlebacked means that the shoulders, back, and flanks are colored, in contrast to the white body. Autosexing dates back one thousand years or more in England and France, even longer in Scandinavia. These breeds probably originated in Scandinavia and are indigenous to areas where Vikings set their anchors. **Shetland** geese, brought to the United States in 1997, may be the original type. They are the smallest of the autosexing geese, and the color patterns in the females vary. Their yellow bills and pink legs hark back to Western Graylag ancestry.

West of England geese, also called **Old English**, may be the geese that the English settlers brought with them when they colonized America. Mixed with Gray geese over time, regional varieties developed, variously called **Cotton Patch** and **Cottonfield** geese. The names reflect their utility on the farm, where the geese were sent into the cotton fields to eat grass and weeds. They are selective eaters, preferring grass to weeds. They carefully avoid crops like cotton and tobacco. **Choctaw geese**, another regional line, may now be extinct.

Females may be either gray or saddlebacked. They may have one or two lobes. Lobes are folds of skin that hang from the abdomen, behind the keel. They range in size from 9 to 11 pounds. Both sexes have pink feet and bills. Other variants in the Gulf of Mexico area have pink shading to orange feet and bills. Dr. Walker suspects they trace their origins back to French settlers.

Pilgrim geese were developed in the 1930s by Oscar Grow. They are a modern version of American Gray

These Pilgrim geese show the difference between ganders (males) and geese (females): the ganders are white and the geese gray or gray saddlebacked. The different plumage makes telling them apart easy, unlike most geese. It's an advantage to be able to distinguish males from females to make decisions for the breeding pen. *Metzer Farms*

geese with the autosexing gene bred into them. Pilgrims have orange bills and legs, which distinguish them from the Old English. They are also slightly larger than Old English geese and are classified as medium at 10 to 14 pounds. They are the only autosexing breed recognized by the APA for exhibition.

CONVENTIONAL (NON-AUTOSEXING) BREEDS

HEAVY

African geese have the recognizable knob on the head, between the eyes. The knob develops to its full size over several years. Although generally males are larger in body size and have larger knobs than females, this is not a reliable way to sex African geese. They vary too much

in size. African geese also have a dewlap, a fold of skin that hangs down under the chin.

African geese are recognized in the Brown color pattern and in solid White. The Browns have black knobs, and the Whites have orange knobs. The Brown is abundant, but the White variety is rare. An unrecognized Buff variety is also raised by fanciers.

African geese thrive even in cold climates, although if they get frostbitten the knobs of the Brown variety may show temporary orange patches that gradually disappear.

The **Embden** (or **Emden**) is the commercial goose that makes its way to the supermarket. Embdens are big, white geese. Occasional gray feathers on young geese usually grow out white as they mature. They grow rapidly to their full size, 16 to 20 pounds for a young goose. Mature geese range from 20 pounds for

These hefty African geese have a knob like Chinese geese do, but, at eighteen to twenty-two pounds, are much larger. As meat birds, they are the leanest of the heavy breeds. The best breeding birds are well-muscled but do not have a keel. The turned-up tail is desirable and a sign of good fertility. *Metzer Farms*

a female to 26 pounds for ganders. This goose feeds a large family at holiday dinners.

Toulouse geese are the dignified ladies and gentlemen of the barnyard. Their patient temperament makes them inclined to gain weight, going beyond the top exhibition weight (26 pounds for ganders, 20 pounds for geese) to 30 pounds or more. The Toulouse is the traditional goose used in France to make pâté de foie gras.

Buff and Gray color varieties are recognized by the APA.

MEDIUM

American Buff geese have the colorful plumage that reflects their name. They are hardy, and their light color makes their carcasses easy to dress out. They were developed from the traditional Gray farm goose and buff-colored geese from Germany.

White Embden geese, which weigh up to twenty-six pounds, contrast with smaller Canada geese, which reach ten to twelve pounds. Embdens are the dominant commercial goose raised in the United States. Canada geese are kept as domestic birds and recognized for showing by the APA *Standard*. Taking geese in the wild is governed by laws, but Canada geese are so numerous that eradication programs are implemented to limit resident populations. *Metzer Farms*

Society for the Preservation of Poultry Antiquities Goose Critical List

African	Old		Pilgrim	Rare
American Buff	Rare		Pomeranian	Old and Rare
China	Old		Roman	Old and Rare
Cotton Patch/	Old and Rare		Sebastopol	Old and Rare
Early American/			Shetland	Old and Rare
Old English/			Steinbacher	Old and Rare
Choctaw			Toulouse	Old and Rare
Egyptian	Old and Rare			
Embden	Old			
Gray	Old and Rare			

SPPA is always looking for established flocks of all rare breeds that have a history of reproducing naturally.

Gray Toulouse geese are a heavy breed that has a dewlap (the flap of skin hanging under the chin) and a keel (a pouch of skin hanging in front of the legs). Their popularity has resulted in development of three separate strains: production, standard, and exhibition. Breeders look for physical size and other qualities, including egg laying. Although their downy feathers keep them warm in cold climates, they have adapted well when introduced to tropical developing countries. *Metzer Farms*

Despite their ornamental appearance, Sebastopol geese are a good utility breed that has a long history in Russia and Europe. They have been called Ripple geese, among other names. The *American Standard* calls for curly feathers all over, including the breast. European birds often have smooth breast feathers and uncurled tails. *Metzer Farms*

Pomeranian geese are a historic German breed, associated with the Pomorze region of eastern Germany between the rivers Oder and Vistula. Although only Gray Saddleback and Buff Saddleback varieties are recognized, they are also raised in Gray, White, and Buff varieties. In Germany, the Buff Pomeranian is known as the Cellar goose.

True Pomeranians are distinguished by their pink bills and pink legs and feet. They have a single lobe. Orange bills and feet or a double lobe indicate that the goose is not a Pomeranian.

Sebastopol geese have long, curling feathers resulting from years of selective breeding. The breed is associated with Eastern Europe, around the Danube River and the Black Sea.

They are a good utility bird, but their novel appearance attracts owners who are inclined to keep them as ornamental birds and as companion birds. They may need the long feathers around their vents clipped in order to breed successfully.

LIGHT

Chinese geese have a knob like the African but are much smaller. Outside the laying season, their slim silhouette has no lobe, which the hens develop at that time. Brown and White varieties are recognized for exhibition. They are the champions of egg laying but don't always make good mothers.

The **Tufted Roman** goose is named for the round tuft of feathers protruding from its head. They have a long European history, going back to Juno's temple in ancient Rome, where they were considered sacred. They originated in the Danube area and are related to Sebastopol geese.

Tufted Romans have a compact body without a keel, lobe, or dewlap, and they make a good roasting bird, despite their relatively small size. The tuft is present from hatching. They are now raised in several colors, although White is the only recognized color.

The eyes are blue, and the bill, legs, and feet may be pinkish or reddish orange.

Brown and White China geese are ornamental and exhibition birds rather than meat production birds. However, those that lay well can be a good addition to breeding programs with lines that need improvement in egg production. For the table, they are leaner than larger geese. A United Nations report recommends them for small flocks. They are alert guardians that will honk loudly at the approach of strangers or predators. Ballantine's Whisky in Dumbarton, Scotland, has a resident flock that guards the warehouse and weeds the grounds. *Metzer Farms*

ORNAMENTAL

Canada geese are the familiar wild birds, but they adapt to captivity well. They breed well but retain some wildness. They may violently defend their pens during spring and summer months. Many subspecies are raised, but the Eastern or Common variety is the one recognized for exhibition.

Their wings must be trimmed annually, after each molt, to keep them from flying away. Clipped wings do not disqualify birds from showing. Although a contented flock will stay put, youngsters may be attracted to fly off with passing wild flocks. This is dangerous to the birds, since they may not have the reactions needed to survive the hunting season.

Egyptian geese are actually Shelducks, a subfamily in the duck, goose, and swan family. They are the smallest of the recognized breeds and the smallest geese raised domestically. Like the Canada goose, wild Egyptian geese still live in their natural habitat south of the Sahara in Africa and in the Nile Valley. They are abundant on the

These White Tufted Roman geese are good small meat birds with moderate laying ability, twenty-five to thirty-five eggs annually. They are calm and easygoing in the poultry yard. The Buff geese are considered a separate breed but the buff color has been bred into others, including Tufted Roman, Sebastopol, Toulouse, and African geese. The buff color variation is a natural color mutation. Breeders bred it into uniformity. American Buff geese have been recognized by the APA since 1947. *Metzer Farms*

Egyptian geese are a separate species from other goose breeds. They are not domesticated in the sense of physically changing as a result of captive breeding. In England and the Netherlands, escaped captive birds kept for ornamental purposes have thrived and established feral populations.
Shutterstock

African continent in areas that provide them with the seeds, leaves, and grasses they eat. Resident flocks have also done well in places where they have been introduced in Great Britain and the Netherlands.

Egyptian geese aggressively defend their territory during breeding and nesting season, so they are best kept isolated from other birds. They may break all eggs not their own and attack other birds and people, even killing other birds.

The recognized variety is beautiful, with a color pattern unlike any other geese: a reddish purple bill, orange eyes, and glossy, iridescent plumage. White and Silver varieties have also been developed.

NON-STANDARD GEESE

Steinbacher geese are a German breed not recognized in the *Standard*, although they have a long history both in Germany and the United States. Steinbacher geese have been fought for sport, as game chickens have been. Steinbachers were standardized as an exhibition breed in Germany after fighting was outlawed.

The **Blue** goose is a strain that has a beautiful blue color. Blue individuals appear in flocks of Gray and Pomeranian geese occasionally. The blue color is regularly bred in Steinbachers. The blue color in geese breeds true, unlike the blue color in other fowl.

Nene Goose

The Nene (pronounced Nay-nay) goose is the state bird of Hawaii. Like many isolated island birds, it is terrestrial, although it has some ability to fly. It migrates from lowlands, where it nests in winter, to higher elevations when it is not breeding. Its migrations are subject to available habitat.

It gets its name from the soft cooing sounds it makes, although it also cackles and trumpets like other geese.

Several factors reduced the Nene from abundance in the eighteenth century to near extinction in the twentieth. The hunting of Nene geese, which was outlawed in 1907, was an initial cause. Nene were not able to recover due to habitat loss and nest destruction. Human development and non-native mammals combined to destroy the plants and nesting sites Nene need. Pigs, rats, cats, dogs, and mongoose preyed directly on Nene geese. Feral cattle, goats, pigs, and sheep destroyed the nesting habitat and food plants they need. By the 1940s, Nene geese had nearly disappeared.

Dedicated individuals rescued a few birds and began a captive breeding program. The Nene was declared a federally endangered species in 1967. Programs to breed Nene in captivity and release them to the wild have succeeded. About 2,500 birds are estimated to live in the wild today. However, only on the island of Kauai is the population increasing without additional captive-bred birds being released. The fact that the mongoose, a major predator of the Nene, has not yet established itself on Kauai is considered the crucial factor. Wildlife biologists continue to monitor Nene flocks closely. Environmental advocates are alert for mongoose, to avoid the destruction that has afflicted the other islands.

The Nene is such an appealing bird, and so friendly to humans, that the public has to be asked not to feed the birds by hand and make pets of them. They quickly become dependent on handouts. They are easily killed by passing cars. Despite early friction, golf courses now welcome them, fencing off areas when a nesting pair moves in. This beautiful bird is beloved by the Hawaiian people. With care, some day it may live in sufficient numbers that it is no longer endangered but a regular visitor and part of the Hawaiian landscape.

Shutterstock

SWANS

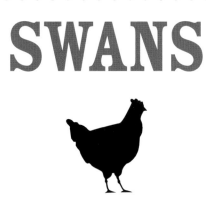

Lakeland, Florida, is known as the City of the Swans. The city acquired its first mated pair of mute swans, the beginning of its breeding stock, from the Royal Flock of England in 1957. Lakeland is actively involved with swan husbandry. Dr. Geoffrey Gardner, the city's swan vet, carries on the traditions of his father, who served as swan vet for more than forty years. The swans are free but content to remain on Lake Morton. *W. Thomas Miles*

The beauty and grace of swans have captured human imagination since before the dawn of history. The fascination continues today, with swans in demand as ornamental waterfowl in parks, country clubs, and golf courses. Lakeland, Florida, received a pair of swans from the Queen of England in 1957 and actively promotes itself as the City of the Swans. Its swan flock now numbers around two hundred birds.

Swans are closely related to geese and, more distantly, to ducks. Scientific classifications are subject to debate and change as new information emerges. Ducks, geese, and swans are grouped together in the Anatidae family, which comprises four subfamilies. Sometimes swans are classified in the same subfamily with geese, Anserinae, but sometimes they are set aside in a separate one on their own, Cygninae. When swans and geese are combined in the Anserinae subfamily, swans are considered the Cygnini tribe. All swans are in the genus *Cygnus* except for the Coscoroba swan, which is presently in a genus of its own. Further research may change that.

Swans are adapted to breeding in cold climates at both ends of the earth. In contrast, true geese are native only to the Northern Hemisphere; the Southern Hemisphere's equivalent waterfowl are the Magpie goose and the Cape Barren goose of Australia and New Zealand, and the Magellan goose of southern South America and the Falkland Islands—none of which are true geese. Geese are terrestrial grazers, whereas swans are adapted to aquatic life, eating more water plants and vegetation, in both fresh and brackish water.

Male swans are called cobs, females are called pens. The etymology for both terms is unclear. Cob may come from the Anglo-Saxon term *copp* meaning "top" (German: *kopf*, "head"). Some attribute it to the Middle English *cobbe*, meaning "leader of a group." Pen may come from the Old English term *pennae*, meaning "feather," or it may refer to the way she carries her wings pinned back. The young are called cygnets.

Swans require three to four years to reach maturity and live about twenty years, so getting involved with raising swans is a long-term project.

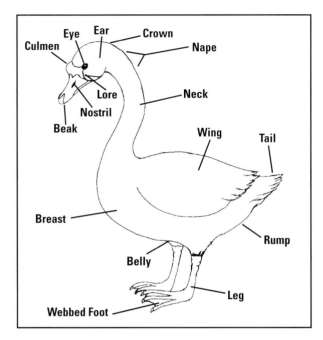

Sharon Wilson

The meat is reputed to be tough and greasy. Most swans are protected by law, but some states allow hunting of Tundra swans. Mute swans, which are not native to North America, are considered an invasive exotic species that is displacing native waterfowl in some places. Some states require farm-raised Mute swans to be pinioned or outlaw them entirely. Because of their beauty and the place this bird holds in our hearts, control measures are controversial. Egg addling and oiling, although labor intensive, are methods of keeping feral populations in check without hunting or otherwise killing swans. An egg is addled by shaking it vigorously for a few seconds, until you hear sloshing inside. To oil eggs, spray corn oil over the top two-thirds of the egg. The oil will seep down and cover the rest, blocking the pores the embryo needs for respiration. A permit is required.

Loss of habitat and environmental pollution are threats to these birds in the wild. Captive breeding populations are an important support for wild species.

HISTORY AND CULTURE

Swans and other birds were hunted by early humans, followed by domestication. Middle Stone Age humans, 8000 to 3000 BC, ate both geese and swans. By the Bronze Age (2000 BC), Mute, Whooper, and Bewick's swans were regular food sources. Swans became semi-domesticated, living close to humans but continuing to migrate. Iron Age Britons revered the swan as supernatural.

Ancient Greeks originated the idea that the swan sang joyfully before it died because, in the words of Socrates, it "could see into an afterlife with the god it served," Apollo. That allusion continues today in the space program's moon landing project. Apollo's chariot was drawn by swans. In the myth *Leda and the Swan*, Zeus takes the form of a swan. The Romans, who had already domesticated geese, ducks, and chickens, may have inaugurated domestication of the migratory Mute swan in England when they arrived in 55 BC.

Traditional Russian shamans in Siberia view the swan as a powerful bird, representing the soul. It is forbidden to kill them and, in some places, dangerous even to point at one or handle a feather. Swan maiden myths are found in cultures worldwide. The essence of this myth is a beautiful maiden who can take the form of a swan. As the stories are told, swans descend from the sky and remove their skins and feathers, emerging as beautiful maidens. One is captured and marries her captor but is never happy. Eventually, she regains her swan skin and flies away to escape back to her home in the sky. Swan maidens combine mortal and supernatural attributes. Russian composer Tchaikovsky wrote the music for the ballet *Swan Lake* about the swan maiden.

In Iceland, before migration was understood, swans were believed to fly to Valhalla, the Norse heavenly Hall of the Slain, or to the moon during winter.

Swans continued to be held sacred throughout Europe through the Middle Ages, the fifth century to the sixteenth century. The landscape of England was marshy during this time—ideal habitat for swans. Swan keeping is documented to the reign of King Edgar in AD 966, prior to the Crusades. Swans were kept in England prior to Richard the Lionheart's Crusade in the twelfth century, from which he is said to have returned bringing swans.

The Swan Knight, Lohengrin, arrived at King Arthur's court in a boat drawn by a swan, which was actually his brother, who had been changed into a swan by his wicked grandmother. Although there is no documentation, later European noblemen claimed ancestral lines back to the

Swans are the largest flying bird. Trumpeter swans have a wingspan of six to seven feet. Tundra swans are smaller. Trumpeter swans are protected throughout their range, but Tundra swans are hunted in North Carolina, Virginia, North Dakota, Montana, Nevada, and Utah.

U.S. Fish and Wildlife Service

Swan Knight. Many family crests from both sides of the English Channel include swans. The Order of the Swan was founded in Germany in 1443. It died out in 1525, but another line was established in the seventeenth century when Mary Eleanor, sister and heir of the Duke of Cleves, married Albert of Brandenburg in Prussia. From that line came Anne of Cleves, fourth wife of Henry VIII. The White Swan pubs in England trace their connections back to her.

Under English law in the seventeenth century, stealing a swan could be punished by transportation to Botany Bay in Australia. As late as 1895, unauthorized possession or killing of a swan carried a penalty as severe as seven years' hard labor.

Swan keeping was a profitable business. Swan keepers were allowed to capture cygnets by royal grant. They would pinion them so that they could not fly,

breed them, and perpetuate them on local waterways. Wild birds gradually became extinct, but the species was secure because of careful stewardship. The practice of keeping swans for the table died out during the twentieth century due to World War II.

Indian legends in South America hold Coscoroba and Black-Necked swans sacred. Coscoroba swans defended the tribe against the Araucanians, the traditional native people of Chile and Argentina. Although they did not succeed, the Sunchild promised to return as a holy child. When his mother was killed before he was born, the unborn baby was rescued by a rat, who raised the child. The rat asked four animals to bring the child to safety. The skunk, who has smelled bad ever since, betrayed the other three animals, but the fourth, the Black-Necked swan, escaped with the child on her back, guarded by Coscoroba swans. Black-Necked swans still carry their young on their backs, as in the legend.

Australian Aborigines see the Black swan as Byahmul, the bird of Byamee, the Great One. A folktale attributes the black feathers and red bill to men who were turned into white swans by their magician brother as a diversion when he and his men raided another camp. The magician then forgot to turn his brothers back into men, and the white swans were attacked by eagles. Crows saved them by pulling out their own feathers and covering the beaten white swans. Their descendants have black feathers from the crows and red bills from their wounds.

SPECIES AND SUBSPECIES

Swans are ancient birds, with fossils showing evidence of them existing as far back as thirty million years. At some point in their development, they divided into Northern and Southern Hemisphere species. Northern Hemisphere species have all white plumage. Southern Hemisphere swans have some or all black feathers. Today, swans are classified into seven species.

If you raise swans native to the United States, they are governed by migratory waterfowl regulations. If you raise non-native swans, they are considered foreign or exotics and may be legally classified as zoo animals. The keepers are responsible for the birds' health and welfare. Although Mute swans were domesticated in Europe for more than one thousand years, no swan species has

Trumpeter swans will breed in captivity but they prefer calm bodies of water that they do not have to share with other swans. The U.S. Fish and Wildlife Service estimated total population, wild and captive, at about 35,000 in 2005. One population of Trumpeters in the wild migrates from breeding grounds in the Yukon to Yellowstone every year. Cygnets have gray plumage that molts to white by the time they are two years old. *Shutterstock*

become truly domesticated in the same way chickens and ducks have. Swans kept domestically do not differ from wild ones.

NORTHERN SWANS

Two species are native to North America, the Trumpeter and the Tundra (also called Whistling). Trumpeters were hunted nearly to extinction in the nineteenth century and remain protected today. Tundra swans are plentiful. Both have been successfully raised in captivity. A Eurasian subspecies of the Tundra swan is the Bewick's (pronounced Buicks) swan. The yellow on their bills is a distinctive marking that varies from individual to individual. All Northern swans are white.

Trumpeter swans, *Cygnus cygnus buccinator*, have long, straight necks and are the largest North American bird

and largest waterfowl worldwide. Males weigh around 26 pounds and measure 57 to 64 inches from head to foot. Females are slightly smaller at 22 pounds and 55 to 60 inches. Trumpeters commonly fly at 60 miles per hour, 80 with a good tailwind. Their legs are set farther forward than those of Mute swans, making it easier for them to walk on land.

The Trumpeter has a strong voice, as its name implies. Anatomically, the swan's windpipe is elongated with an upward loop to the back. A partition of bone separates a section of the windpipe's fold within the breastbone. They use a loud call during migration to keep the flock together. In family groups, they communicate conversationally and give warning and decoy calls.

Tundra or **Whistling** swans, *Cygnus columbianus columbianus*, are similar to Trumpeter swans in plumage and

Tundra swans are the most widespread and numerous species of swan in North America. There are two distinct populations in the East and the West. The eastern population winters along the Atlantic coast and breeds in the Arctic during the summer. The western population winters in California and Utah and breeds along the west coast of Alaska and the Arctic. The current population is estimated at around 160,000 birds. As migratory birds, they fly more than 3,700 miles twice each year. *Shutterstock*

color but smaller, with a maximum weight of 18 pounds for a body that is 45 inches long with a 62-inch wingspan. Its shorter windpipe gives it a different voice from the Trumpeter. Its calls include a high-pitched whistling, soft laughter, and a high-pitched whistling bark. It may fly as fast as 100 miles per hour on migrations but locally makes short, low flights.

Tundra swans take their name from the cold arctic tundra where they breed, but they have been successfully raised in warm climates.

Whooper swans, *Cygnus cygnus cygnus*, are the European and Asian variety of Trumpeter swans. The Whooper is similar in appearance and size to the Trumpeter, 55 to 61 inches long with an 81-inch to 93-inch wingspan and weighing from 17 to 33 pounds.

The windpipe is not as long as the Trumpeter's, nor is it divided with a bone partition, but it does have a small loop. That gives Whoopers the ability to vocalize loudly, with buglelike calls to the group and softer, high-pitched vocalizations for conversation. Although it is a cold-climate bird in the wild, captive Whoopers have been successfully bred in warm climates. It is the national bird of Finland, and its image honors the Finnish one-Euro coin.

Bewick's swan, *Cygnus columbianus bewickii*, is native to the northern reaches of the Arctic Circle. Usually considered a subspecies of the Tundra swan, it is about a third smaller. Its bill also has more yellow and is straighter than the Tundra's. The yellow bill markings are unique to individuals.

Measurements of Trumpeter, Tundra, and Mute Swans

	Trumpeter			Tundra		Mute		
	mean		range	mean (sd)		mean		range
	male	female	male & female	male	female	male	female	male & female
Weight (kg)	11.9	9.6	7–14.5	7.2 (0.8)	6.3 (0.7)	10.2	8.4	6.1–14.1
Total length (mm)	1,477	1,435	1,398–2,578	Range 1,200–1,470 both sexes		—	1,410	1,270-2,520
Wingspan (mm)	—	—	1,890–2,450	1680 both sexes		—	—	2,080–2,380
Bill length (mm)	120	116	101–231	104 (5.2)	101 (5.1)	103	98	92–218
Egg length (mm)	117	—	101–224	106 (range 96–216)		103	—	113
Egg weight (g)	363	—	239–410	273 (range 210–340)		295	—	258–365

Shutterstock

All data except wingspan obtained from *Birds of North America* species accounts (Nos. 105, 89, and 273). Additional data from the Cornell Lab of Ornithology and the British Trust for Ornithology. *The Trumpeter Swan Society*

The **Jankowski** swan, *Cygnus columbianus jankowskii*, is another Tundra subspecies sometimes called the Eastern Bewick's swan. It breeds in the Lena delta area of eastern Russia.

Trumpeter and Tundra swans, whose habitats overlap in places, are able to interbreed and produce fertile offspring. They are called **Trumplers**. Crosses between Trumpeters or Tundras and Mute swans are sterile.

Mute swans, *Cygnus olor*, which do make some sounds, are native to Europe and Asia. The species' long history in English lore has Richard the Lionheart introducing them to England on his return from the Crusades, but historical documentation for swans in England predates the Crusades. The earliest recorded statement of ownership is from AD 966, when King Edgar gave rights over stray swans to the abbots of Croyland.

Mute swans are the classic swans, with their S-shaped necks. Full-grown birds measure 50 to 60 inches from bill to tail and weigh 18 to 35 pounds. They are the heaviest of all flying birds and have an average wingspan of 6 to 7 feet. Record size and wingspan are much greater, with the largest specimens topping out as heavy as 50 pounds and having a 12-foot wingspan.

Black swans from Australia are in demand as ornamental birds at hotels, golf courses, and country clubs. They will breed in captivity but can be aggressive. Facilities should provide separation to avoid injuries to visitors as well as to protect the birds from predators. *Shutterstock*

The Mute swan is graceful on water, with its large webbed feet—6½ inches from leg to toe—functioning as both propeller and rudder. It has a protrusion at the top of its bill called a knob.

The traditional English color variety has a reddish-orange bill and black legs and feet. The cygnets are gray. A leucistic, or reduced pigmentation, variety is called Polish. It has an orange bill and buff legs and feet. The cygnets are white.

SOUTHERN SWANS

Southern Hemisphere swans include the Australian Black swan, the South American Black-Necked swan, and the South American Coscoroba swan. Southern swans are the only ones that have any black plumage.

Australian **Black** swans, *Cygnus atratus*, range in size from 43 to 55 inches long, have a 63-inch to 78-inch wingspan, and weigh 13 to 20 pounds. Their feathers are not truly black but have a brownish tinge. The feathers on the back have a gray border, and the primaries are white. White and Brown varieties of Black swans have been developed through selective breeding, and established strains are available. The bills and eyes are red, but the male's eyes turn white during breeding season.

The Black swan has a shorter windpipe that makes the bird quiet relative to the loud Northern swan but

Black-Necked swans are the only swans that have a caruncle on the beak. The male's is often larger and redder than the female's. They become quite tame and are admired ornamental waterfowl. Their short legs are farther back on their bodies than the Coscoroba, with whom they share most of their range; this characteristic makes them clumsy on land. Coscorobas' legs are longer and located more forward, and they walk easily on land. *Shutterstock*

more vocal than the Mute swan. It has a high-pitched, musical bugle.

The species is adaptable to a wide range of climate conditions, which has allowed its population to recover in the wild after being threatened by hunting and habitat destruction in its native Australia.

The two South American swan species overlap in their habitats. They have developed differently, perhaps evolving to flourish together in separate ecological niches. The Black-Necked swan is more aquatic, making greater use of water resources, while the Coscoroba is more terrestrial, making use of the land.

The **Black-Necked** swan, *Cygnus melancoryphus*, is a species distinct from its northern relatives. It has a blue bill with a red caruncle at its base. Adults range from 45 to 55 inches long. Males are significantly larger than females, weighing 12 pounds to her 9 pounds.

Anatomically, the Black-Necked swan's pinkish legs are farther back on its body, making the bird awkward on land. The windpipe has only a slight bend, and the swan's vocalizations are a high-pitched whistle and a low honk. It has relatively short wings but is a powerful flier.

Their native habitat takes Black-Necked swans from the warm tropical climates of Argentina to as far south as

The face and lores of the Coscoroba swan are fully feathered, unlike in other swans. The breed resembles whistling ducks in some respects, such as the shape of its crimson bill, but it remains classified as a swan. The Coscoroba is an omnivorous dabbler that eats leaves and stems of aquatic plants, insects, fish spawn, and small crustaceans. This individual is either a juvenile or a female, which have brown irises. Mature males have whitish eyes. *Shutterstock*

the extreme weather conditions of the Falkland Islands, but they can be delicate in captivity. Nevertheless, captive populations have bred successfully and are prized as unusual ornamental birds.

South America's small **Coscoroba** swan, *Coscoroba coscoroba*, is gooselike in appearance and is in some ways similar to the Whistling ducks. Whistling ducks are a separate genus (scientific classification) of waterfowl more closely related to geese and swans than to "true" ducks. They have long legs and necks, and males and females are similar in plumage. There are only eight whistling duck species in the world, two in the United States, all wild. This swan's trumpeting call—*cos-co-ro-ba*—gives the bird

its name. It is small, 35 to 45 inches long and weighing 7 to 11 pounds. It has a carmine-crimson bill and legs, which are long and set forward, allowing it to walk well on land. Its black wing-feather tips constitute the only black in its plumage.

Sylvia Bruce Wilmore in *Swans of the World* considers the Coscoroba "the droll, endearing clown of the swan family" for its odd behavior. It moves its head back and forth as some ducks do and feeds by spreading its wings and turning end-up.

Breeding success in captivity is limited.

FEEDING

Like geese, swans are grazers but do much of their grazing on water plants. Favorite foods include eelgrass, common sedge, amphibious bistort, bur-reed, Canadian pondweed, and reedmace. Captive swans can be fed from feeders on poles in the water to keep feed away from other birds and wildlife. Layer pellets mixed half-and-half with cracked corn or grain provides a good supplemental feed. Turkey and game-bird feeds may be too high in protein and cause angel wing. Feed must be kept clean and dry. Check it every other day, or every day in wet weather. If it is wet, empty it, wash it, and dry it completely before refilling and replacing.

Swans will enjoy eating the landscaping, so it's wise to plan on allowing them to enjoy it. Make sure the plants they have access to are not toxic to them.

BREEDING AND RAISING CYGNETS

Swans lay fewer eggs than other waterfowl, from two to twelve, usually around six per breeding season. Each egg measures about 4 ½ inches long and weighs 12 ounces. Incubation lasts around thirty-five days. If the nest is destroyed, the pen may lay a second, smaller clutch. It is normal for the pen to lose weight while incubating. Good food supplies can help her store fat before incubation begins.

Swans hatch their eggs better than artificial incubators do. The cob chooses a nesting site, but if the pen doesn't like it, she may reject it, and he will have to find another more to her liking. The cob brings nesting materials to the pen, who arranges them. She places the larger roots, rushes, and sticks on the outside and lines the nest

This swan feeder is actually a dog feeder, mounted on a pole in water at least two feet deep. That keeps it secure from land animals. Posting it at least fifteen inches from the surface makes it inaccessible to ducks and geese. Assuring swans of a secure food supply will keep them faithful to the location, even if they are frightened away by high winds or other extreme weather.
W. Thomas Miles

with decaying vegetation and some of her own down. If she did not remove down from her breast, her feathers would insulate so effectively that the eggs would not be warmed. Wilmore describes a setting pen as "a snow-queen on a very untidy throne."

Both parents rear the young together. The parents molt at the end of July, while they are raising their cygnets. The pen molts first, followed by the cob several weeks later. Thus, only one parent at a time is vulnerable. They lose all flight feathers during the molt and are flightless until new feathers grow in.

To increase swan production, Muscovy ducks and geese can be recruited as foster parents for the first clutch, leaving the pen to raise her second clutch. Swan eggs are tricky to hatch artificially, and the amount of attention cygnets require is prohibitive for most humans. Imprinting is a pitfall, since the cygnets will identify with their first keeper. In the case of ducks and geese, this may result in failure ever to mate successfully with another swan. In the case of humans, it may expose the swan to danger if it becomes too trusting of humans. On the flip side, swans can serve as surrogate parents to goslings. SPPA president Craig Russell observed a family of swans that included a gosling on the Nahe River in Germany in the 1970s. The cygnets were about three days old. The gosling was part of the swan family, sailing up and down the river, very much part of the flock.

Swans hand-raised from cygnets will learn their names and respond to their keepers. A close relationship with their keepers makes catching them for veterinary care

Among swans, Bewick's eggs are the largest and Mutes the smallest. Because they lay so few eggs, practically speaking each egg has its greatest value as a potential bird. Where feral populations of Mute Swans cause problems, "addling" their eggs (shaking them so that they will not develop) or coating them with oil to prevent development are sometimes employed as non-lethal ways to reducing population. *Shutterstock*

Cold Feet

The feet of swans and other waterfowl always feel cold because the blood flow to the legs and feet is reduced in temperature through a physiological adaptation. This adaptation conserves body heat from these unfeathered parts of the body. It helps the birds keep their body temperature at safe levels even when their swimming is the only thing keeping the water from freezing over. Average body temperature is 106 degrees Fahrenheit.

and other maintenance much easier. Hand-feeding can be risky, however. They can get excited and assertive in demanding food, to the fright and injury of their feeders.

HEALTH MANAGEMENT

Swans are subject to avian diseases that affect other birds, including avian influenza, duck virus enteritis (duck plague), aspergillosis, Newcastle disease, and Marek's disease. They also can contract botulism. The bacteria that produce the neurotoxin, *Clostridium botulinum*, can grow in the anaerobic environment at the bottom of the pond. Swans stir up the bottom to feed on the grubs and worms that live in it and inadvertently ingest the bacteria. The toxin causes paralysis, beginning at the feet and progressing to the wings and neck. Botulism is also known as limberneck. Swans, geese, and ducks can all be affected and may die within hours of eating contaminated items. Death follows convulsions and paralysis of breathing muscles.

An effective vaccine exists, given as a series of four injections. However, birds may be subject to botulism before the full series has developed immunity in the bird.

Their feet may develop infections, leading to bumblefoot. If left untreated, such infections can cause lameness and eventually death. The infections may need to be surgically cleaned by a veterinarian. Both systemic oral antibiotics and local applications may be necessary to cure the infection.

The Lakeland swan keepers suggest donuts as a treat to camouflage medicines. Avoid indulging in such sweet treats unless needed for medical treatments.

HOUSING

Swans mate for life, although around 15 percent may break that bond at some time during their lives. Wilmore says, "There appears to be a real attraction between particular individuals of the opposite sex, and not just a pair-bond associated with the sexual life: they stay together outside the mating season and apparently enjoy each other's company." Mute swans do not migrate much, and a mated pair will remain in the same nesting territory year-round. Breeding swans are territorial and may act aggressively. They should be kept separate from other birds. Consider that a single pair of Trumpeter swans in the wild may occupy an eight hundred-acre lake.

Allow your swans to settle into a new home for three to six weeks before giving them freedom to explore. Introduce them to the new location and give other resident swans an opportunity to get acquainted from behind fencing.

The fenced-off enclosure should include part of the water of their new home. They require water to swim in. They eat, bathe, and mate in the water. After the initial holding period, the enclosure can be opened on the water side. Continue to maintain the feeders and any shelters that the birds have become accustomed to until they have safely transitioned to the new location.

Swans in public settings may appreciate a private area away from public view. At Orange Lake Resort & Country Club in Orlando, Florida, the swans have their own retreat area. Protected by a fence lined with thorny bougainvillea on one side and a fence with a viburnum hedge on the other, the space discourages intruders both human and animal from entering and gives swans advance warning of any approach. The retreat gives swans a place of their own so that they stay away from public areas where they could become a nuisance, such as the golf course. The swans feed, preen, socialize, and care for their cygnets there.

Mute swans have been valued domestic birds for hundreds of years. Although most live wild now, they retain a semi-domesticated nature. In England, swan marks, made by cutting the bill so that it is permanently scarred, functioned much like cattle brands. Leg and foot markings were also used. *W. Thomas Miles*

Feather Care

Feathers are important. Swans bathe at least once daily, in a dramatic demonstration of splashing. They enjoy drying their feathers after getting out of the water. They flap their wings to get their feathers straightened out. They preen them, nibbling at the uropygial gland at the base of their tail and spreading waxy oil on their feathers. Whether this is done to waterproof the feathers or add vitamin D to them remains under investigation.

W. Thomas Miles

SPECIALIZED CARE

Lakeland's swan keepers have collected their knowledge and experience with swans into a book, *The Swan Keeper's Handbook*. They address management issues from the perspective of their civic history with swans.

Swans are powerful flyers, sustaining speeds around 30 miles per hour but ranging from 18 to 55 miles per hour. The long, broad wings of Mute swans may flap 160 times a minute in a figure-eight stroke. On local flights, they may fly at heights of between 50 and 100 feet, but migrating swans climb to 2,000 and 5,000 feet, occasionally flying as high as 10,000 feet.

Captive swans are usually pinioned to prevent them from flying away. In the case of non-indigenous species, this measure keeps them in the location for which they are intended. In New Hampshire, Mute swans are required to be pinioned. Swans kept in populated areas need to be confined to avoid collisions with power lines and vehicular traffic. In Florida, they could land on water occupied by alligators and be subject to predation. The pinioning operation is recommended at three weeks of age, before the blood supply to the wing is fully developed. If birds are not pinioned, their wing feathers need to be clipped every six to eight months to keep them from flying because the feathers will molt and grow back.

Swans are strong, and reports of swans breaking a human's leg are not exaggerated. They can also scratch with their claws and bite with their bills. Either can draw blood and cause a nasty wound. As the swan keeper, you need to be the dominant "bird" in the group, without being drawn into a fight. Always be aware of the potential for aggression. Stand tall and give the appearance of a larger bird, by waving your arms or a white towel. Back away when necessary. All swans are territorial. Any bird can be dangerous if threatened or protecting a nest or cygnets.

Swans capture our attention wherever we see them. However, they require special care and should not be acquired without study and apprenticeship under an experienced caretaker. Learning about these beautiful birds can only increase your love for them.

TURKEYS

Turkeys are a great choice for small flocks. They have become good business for industrial flocks, but they deserve better.

Because they can tolerate the crowded conditions that maximize return on financial investment, they have become a dominant domestic bird raised in the United States. The commercial Broad Breasted White birds raised in those conditions are not recommended for small flocks, though. Commercial turkeys have been selected for breast meat development to such an extreme that the toms are physically incapable of mounting the hens or making the contact required for successful fertilization. The qualities that work in favor of industrial success are detrimental to small-flock raisers.

Concentrated Animal Feeding Operation (CAFO) practices made it possible to raise 262,460,000 turkeys in the United States in 2006. They provided inexpensive meat for a growing market eager to purchase it. CAFOs run the risk of incubating highly pathogenic organisms along with the birds. Most consumers familiar with the living conditions find them distasteful, even repugnant. Public awareness of the subject is limited, although it occasionally makes headlines, as in the case of inhumane treatment of downer cows in 2008. Conversely, raising livestock in small flocks makes humane conditions possible. Consumer awareness has created a niche market for table-ready turkeys.

"Centuries of domestication have not changed the turkey's natural love for a necessity of free range," wrote Herbert Myrick a century ago in *Turkeys and How to*

Industrial operations artificially inseminate turkeys. At the ARS Germplasm and Gamete Physiology Laboratory in Beltsville, Maryland, poultry physiologist Ann Donoghue candles turkey eggs to identify fertile ones. *Keith Weller*

Grow Them. That remains true for traditional breeds. Although they can be successfully raised in turkey "porches" and yards, they do best when they can have range or pasture on which to forage. They are valued as pest-eating protectors for crops. In the eighteenth and nineteenth centuries, turkeys were driven long distances to market without supplemental feed, eating what they found along the way. Today's turkeys retain much of their wild character, although they become quite tame when raised with human attention.

HISTORY AND CULTURE

Turkeys are native to the Americas, a unique contribution to our tables. They were domesticated in Mexico and Central America for centuries before European contact.

The Ocellated turkey is a species of wild turkey native to the Yucatan peninsula in Central America. Its name refers to the eye-shaped bronze spots on their tail feathers, which are bluish gray and tipped with gold. Early scientists thought these birds might be related to peacocks, but they are not. They are smaller than North American turkeys, with toms weighing no more than twelve pounds and hens no more than eight. *Shutterstock*

The Aztecs called the turkey *huexolotl*, which sounds like a turkey's gobble, and deified it as Chalchiuhtotolin, meaning the "jeweled bird." They raised turkeys as domesticated birds by the thousands. Because the turkey was important to Native American cultures, it held a place in many tribes' mythology. Navajos told how Turkey helped re-create the world after a flood. Turkey taught the Apache how to raise corn. Cherokees have tales about how the turkey got its wattle and why it gobbles. Santa Clara Pueblo and Zuni people explain why turkeys scatter when they hear humans with a tale of betrayal by the girl who kept them. The Tewa Pueblo Indians tell about the turkey as a food source.

Turkeys continue to capture our imagination. Ted Andrews attributes to the "earth eagle" the bounty of "all the blessings the Earth contains, along with the ability to use them to their greatest advantage."

The emperor Montezuma and his court raised turkeys for their own consumption and to feed the carnivores and raptors in his zoo. W. H. Prescott in his *History of the Conquest of Mexico* (1844) writes that the court feasted on eight thousand turkeys during a single year.

Turkeys were one of the riches European explorers brought back to their home countries. Turkeys became

The Codex Borbonicus dates to 1507, prior to Spanish conquest in 1518 to 1521. It was composed by Aztec high priests, summarizing the religion. It allocated portions of the calendar year to dominant gods and goddesses like Chalchiuhtotolin, The Jeweled Turkey. The Codex was used to propitiate deities and direct personal and civil events under auspicious auguries. *Sabine Eiche*

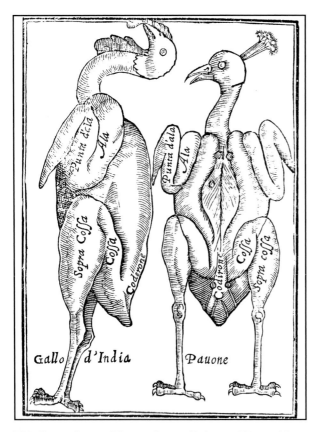

This illustration on "How to Carve a Turkey and Peacock" comes from Vincenzo Cervio's *Il trinciante*, 1594. The turkey carver provided part of the entertainment at banquets in sixteenth-century Rome. Spearing the bird with a long fork from back through breastbone, the carver held it high in the air over the platter and sliced off pieces that then fell into an elegant arrangement on the platter, "with the utmost grace and dexterity." *Sabine Eiche*

very popular in Europe. In a letter to his chief treasurer in the West Indies in 1511, King Ferdinand of Spain ordered every boat sailing back to Spain to include five male and five female turkeys "so that they start a breed here." They were enthusiastically received and soon were celebrated in heraldic crests as well as on the festive table.

Europeans thought they were a kind of peafowl. In England, the turkey was confused with guinea fowl, which was already established there, and for a time the name "turkey" could apply to either. Linnaeus, the founder of the modern animal and plant classification system, compounded the confusion by giving the American turkey the name *Meleagris*, the traditional Greek name for "guinea fowl." The scientific name for turkey is *Meleagris gallipavo*, with the species name coming from *gallus* for "chicken" and *pavo* for "peafowl."

Later European colonists brought domestic turkeys back to North America with them. The domesticated turkeys had been bred to larger size, a typical change among domesticated birds due to selective breeding, more food, and less exercise. Being uniquely American was one of the arguments Benjamin Franklin made in favor of making the turkey the national bird instead of the eagle. "For in Truth, the Turk'y is in comparison a much more respectable Bird, and withal a true original Native of America. . . . He is, [though a little vain and silly, it is true, but not the worse for that,] a Bird of Courage, and would not hesitate to attack a Grenadier of the British Guards, who should presume to invade his Farm Yards with a red coat on," he wrote in 1784.

William Bradford, governor of Plimouth Colony, recorded some years after the event that turkey had been part of the first Thanksgiving feast in 1621. He also mentioned "water foul," which would have included wild geese, several kinds of ducks, and swans.

Wild turkeys were hunted to extinction in their traditional ranges as settlers cleared woodlands and populations moved west. Flocks of wild turkeys that once numbered in the hundreds were thinning as early as the 1670s.

Market hunters became wasteful because of the abundance. In the early nineteenth century, whole wagon-loads of turkeys spoiled due to lack of refrigeration. Hunters often shot birds indiscriminately and took only the ones that weighed more than 15 pounds, leaving the rest to rot. Some preserved only the breast and discarded the rest.

By 1813, the last wild turkey was seen in Connecticut. Moving west, wild turkeys were gone from New York by 1844, Kansas by 1871, and Iowa by 1907. As hunters shot without discrimination and the land was deforested, turkeys in the wild reached a low in the late nineteenth and early twentieth centuries. One estimate puts the total wild turkey population at thirty thousand around the turn of the nineteenth century.

As the number of wild turkeys plummeted in the nineteenth century, domesticated birds thrived on farms.

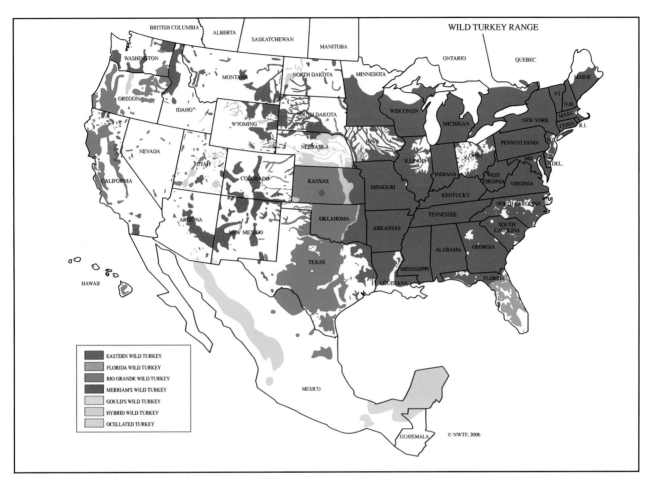

WILD TURKEY RANGE

EASTERN WILD TURKEY
FLORIDA WILD TURKEY
RIO GRANDE WILD TURKEY
MERRIAM'S WILD TURKEY
GOULD'S WILD TURKEY
HYBRID WILD TURKEY
OCELLATED TURKEY

© NWTF, 2006

Turkeys originally ranged widely across North America and were an important part of Native American and Aztec cultures. Colonists and settlers reported seeing flocks of hundreds of birds. James Audubon wrote in 1900 that he saw "What would be said to be a gang of Wild Turkeys—several hundred trotting along a sand-bar of the Upper Mississippi."
National Wild Turkey Federation

They were common enough for Washington Irving to note in *The Legend of Sleepy Hollow*, published in 1820, that Ichabod Crane saw "regiments of turkeys were gobbling through the farm-yard" on the estate of Balthus Van Tassel, father of the beauteous Katrina.

The average nineteenth-century farm flock consisted of four to twelve hens and one to three toms, providing enough birds for the family's table and a few to sell. Farmers bred for color and size, developing color varieties that would eventually be recognized in the American Poultry Association's official *Standard of Excellence*.

Government programs for wildlife habitat restoration and private organizations like the National Wild Turkey Federation have helped wild turkey populations

recover. Around seven million turkeys were pursued by three million hunters in the United States in 2008.

Renewed interest in heritage breeds has increased the market for these specialty birds. The steady growth of demand that exceeds supply indicates that heritage turkeys, such as Bronze, Narragansett, and White Holland, can present small-flock owners with a reliable source of income.

VARIETIES

Turkeys vary a lot in size, so that's a consideration in selecting the kind to raise. Plan on adequate space for the breed you select. The largest turkeys are adult toms of commercial breeds, which can weigh more than 36

The Narragansett turkey is one of the heritage breeds being raised to meet the demands of the growing market for them as holiday birds. They take longer to mature to market weight than commercial turkeys, but the time allows the meat to develop more flavor. Cooks and diners also find the meat firmer than that of commercial turkeys. *Robert Gibson*

Frank Reese Certified-Humane Good Shepherd Ranch

Frank Bob Reese has led the movement to return *Standard*-bred turkeys to America's table. Even as a child, he was fascinated with turkeys. As an adult, his apprenticeship with Norman Kardosh—national leader in turkey breeding, for whom the Kardosh Bronze was named—ended only with Kardosh's death in 2004. Reese promised him that he would preserve historic turkeys.

"That's one of the meanings of 'Good Shepherd Ranch,'" he says. "I intend to be a good shepherd to these birds."

His father counseled him to get a career that would provide income, which he did. He is an anesthesia nurse employed full time at a local hospital. Beyond that, for more than forty years he has devoted his attention to *Standard* Bronze, Bourbon Red, Buff, Narragansett, and Beltsville Small White turkeys at his family's 160-acre ranch in Lindsborg,

Kansas. The turkeys are raised on pasture, guarded by four large dogs.

"The best way to save the old-time poultry is to return them to our dining tables," he says. He developed a partnership with Heritage Foods USA to market them. Slow Food USA includes Reese's heritage turkeys in its Presidia. Increased demand compelled him to partner with other producers, who all adhere to his pasture growing regimen. Good Shepherd Ranch was the first turkey ranch to receive the Animal Welfare Institute's certificate of approval.

Reese sells more than ten thousand turkeys a year. His leadership has raised awareness of historic livestock breed conservation, particularly of poultry. He is currently focusing on creating a Standard-Bred Poultry Institute to educate and advocate for husbandry of *Standard*-bred poultry breeds.

pounds; the smallest specialty breeds top out at 10 pounds for hens. Generally, the range is 12 to 33 pounds.

Large White and Medium White commercial poults (baby turkeys) are available for small-flock raisers. White turkeys have the advantage of presenting a clean carcass because the white pinfeathers don't show in the white skin. However, White turkeys actually have more feathers than breeds with darker plumage. White Holland is the name given to commercial white turkeys, but they are not the same as the historic White Hollands. They have been selected from Broad Breasted Bronze flocks and bred for a larger, meatier breast. The *Standard* changed the weight requirements in 1983 to reflect the larger birds being grown under this name, making them the same as for Bronze turkeys: 36 pounds for an old tom and 20 pounds for an old hen.

Commercial Broad Breasted or Large Whites mature as early as twelve weeks for hens, not later than eighteen

weeks for toms. To market smaller table-ready birds, they may be processed as early as eight weeks. Smaller varieties like the Midget White have been developed, but they are rarely raised now. The market for smaller birds is served by slaughtering younger Large Whites instead, since they develop the characteristic large breast early on.

Historic breeds are a great choice for your own flock. If you are going to put in the time and space for turkeys, consider raising varieties that reward you for the effort with more than simple conversion of feed into meat.

Heirloom breeds take longer to mature to table size, typically six to eight months. Reaching full size and maturity may take a year.

Unlike ducks and geese, all domestic turkeys are the same breed. They are all derived from wild turkeys, which have in their genes all the colors we see in the many recognized and unrecognized color varieties. With

White turkeys date back to the time of the Aztecs. The White Holland shown here is of the variety that led to the modern industrial Broad Breasted White. Although all large white turkeys are often lumped together, the historic strain is distinct from the industrial strain and has longer legs and less pronounced breast development. These birds can mate naturally. *Shutterstock*

so many colors expressed in the plumage, the imagination of turkey lovers has reached to find names for them all, including Cinnamon, Nutmeg, and Chestnut. In unrecognized varieties, some confusion exists, with similar names applied to variations on the color theme. Think of color varieties in turkeys as branches on a bush, overlapping and sharing qualities but relating to the original wild colors.

White Holland turkeys are descended from white farm turkeys that were developed through selective breeding in the Netherlands and Austria. Their advocates husbanded and bred them in the eighteenth century, despite a popular prejudice against white plumage at the time. The belief that any white bird was constitutionally weak may have developed from observed weaknesses in flocks that were inbred. It may have come from the fact that, on open range, white birds are more subject to predation than birds that have more camouflaged coloration. The limited open pasture of the Netherlands reduced predation and made white birds more practical. The White Holland variety was brought to the United States in the early nineteenth century and was included in the first APA *Standard*

of 1874. Although they are recognized as commercial turkeys by the APA, strains of the original birds are distinct and still raised by fanciers. The variety originally had blue eyes, but brown eyes are now specified in the *Standard*. The historic strain of White Hollands should weigh the same as Narragansetts and Blacks: 33 pounds for old toms and 18 pounds for old hens.

Breeders worked to develop both larger and smaller White varieties. Selective breeding made White Hollands the basis of the American Beltsville Small White turkey and today's Broad Breasted White commercial turkey. Beltsville Small White turkeys were developed during the 1940s at the USDA's Research Center in Beltsville, Maryland. The variety was admitted to the *Standard* in 1951, with old toms weighing 21 pounds, old hens 12 pounds, young toms 17 pounds, and young hens 10 pounds.

Midget Whites were developed from the Beltsville birds, topping out with old tom weights of 13 pounds and old hens at 9 pounds. They grow a large breast and have conformation similar to that of Commercial Broad Breasted Whites but not developed to such an extreme. Because they are smaller overall, they are able

Laying Eggs

Although turkey hens lay a greater number of eggs during their first year, fertility is better in two- and three-year-old traditional-breed hens. They may start laying as young as thirty weeks old. Hens that are younger than that in the spring of their first year may lay smaller and fewer eggs. "I always keep older hens with younger toms," said experienced breeder Frank Reese of Good Shepherd Turkey Ranch in Kansas.

Turkey eggs typically weigh 2¼ ounces to more than 3½ ounces. Some lines of turkeys selected for egg production for forty-three generations now lay an average of 198 eggs annually, but their eggs are closer to the size of chicken eggs.

Gather eggs two or three times a day to encourage laying, Reese advises. He gets sixty to eighty eggs a season from his Bronze hens.

Midget White turkeys were bred for egg laying, and sixty to eighty eggs a year is typical. Using artificial light to lengthen the laying season can extend laying for other varieties to as many as one hundred eggs a year.

Slow Food USA has experimented with turkey eggs in cooking and baking. The French Culinary Institute volunteered its facilities for the use of Yuri Asano of Slow Food USA; Tina Casaceli, director of Pastry Arts and Baking Arts at the Institute; and two FCI students, Ji Won Kim and Yo Ok Kim.

Their conclusions were that turkey eggs are excellent in dishes with thick, rich texture and taste, such as crème pattisiere (pastry crème), crème brûlée, crème caramel, and deviled eggs. They are less successful in lighter dishes like angel food cake and Génoise cake.

Shutterstock

to mate naturally. Midget Whites were popular in the 1950s and 1960s.

Although the demand continues for small turkeys, these varieties no longer have a commercial following. Some breeders maintain flocks, and they can be a good choice for small flocks. In the past, they have been selected for egg production, laying sixty to eighty eggs annually. This could be a focus of a breeding operation.

The **Bronze** is what most of us think of as a classic American turkey. Breeders have nurtured several color varieties in this pattern, which harks back to the original wild turkey plumage. The domesticated *Standard*-recognized bird is larger than its wild ancestor, is lighter in color, and has white tips on the tail feathers, where North American wild turkeys have black. A Mexican wild variety has white-tipped feathers.

The Broad Breasted Bronze was developed from the historic *Standard*-recognized Bronze to meet market demands for larger carcasses that have more white meat. Those who want to preserve the historic variety should ascertain that they are getting *Standard* Bronze stock, not Broad Breasted Bronze. Confusingly, Broad Breasted birds are often the variety exhibited, although judges are sensitive to the distinction. On the other hand, adding a few Broad Breasted hens to a breeding program of *Standard* or wild turkeys will increase the meat on their

Bronze turkeys forage for their food at Mike Walters' free-range turkey ranch in Oklahoma. Traditional breeds benefit from hunting for their own grasshoppers and other insects, acorns, and seeds on pasture. In the eighteenth and nineteenth centuries, farmers would herd their turkeys to market in a Turkey Trot. They would bring some grain with them to keep the flock together, but the turkeys foraged and fed themselves on the trip. *Mike Walters*

offspring without hurting fertility. Hens resulting from that crossing will be more inclined to be capable brooders and mothers.

The Broad Breasted Bronze was developed in England in the 1920s and imported to the United States in the 1930s. Selective breeding has increased its breast size even further, making natural breeding impossible. It became the dominant commercial breed after World War II. It was supplanted by the Broad Breasted White in the 1960s, again meeting a market demand for a cleaner carcass because of the white feathers.

Bronze variants include the Crimson Dawn, also called the White-Winged Bronze; the Black-Winged Bronze; the Dark Brown; the Light Brown or Auburn; and the Silver Auburn. Crimson Dawn turkeys, whose feathers have a pink tinge, have white shoulders (hence White-Winged) and black-tipped flight feathers. Black-Winged

Bronze turkeys are a black version of wild turkey coloration and may be one of the early variations of domestication. The Dark Brown has brown where the *Standard* color pattern has black. A redder version is the Copper. Light Brown has reddish brown (hence Auburn) where the *Standard* has black, and it has tan where the *Standard* has bronze. The Silver Auburn has white instead of tan.

The Bronze variants, which are not APA-recognized, tend to be slightly smaller, like the Narragansetts and other varieties, than the *Standard* Bronze. Broad Breasted Bronze turkeys, especially toms, may grow much larger.

The **Narragansett** is the historic turkey of New England. It takes its name from the Narragansett Bay of Rhode Island and was included in the first American *Standard of Perfection* in 1874. The color pattern is similar to that of the Bronze, with steel gray or dull black in place of bronze and tan instead of brown in the tail. If

The contrastive plumage of Royal Palm turkeys makes them visually striking. American records trace this variety to Enoch Carson's flock in Lake Worth, Florida, in the 1920s, but birds with similar plumage are documented in Europe as far back as the eighteenth century, where they were called Pied, Crollwitzer, or Black-Laced White turkeys. *Christina Tyzzer*

it didn't have such a long history, it would be included as a variant of Bronze, but the two have been separately husbanded for hundreds of years.

Silver Narragansetts, with white in place of the usual steel gray and tan, occur naturally in flocks. They can be selected and bred into a distinct strain but are very rare.

The **Royal Palm**'s distinctive white-and-black plumage makes this small turkey, just 10 to 22 pounds, striking in appearance. Its pattern is similar to that of the Narragansett. Its white feathers are edged with black, and its solid white body contrasts dramatically with black saddle feathers. It was developed in the twentieth century as an ornamental bird, although records of birds with such feathering date back to the 1700s. The pattern appeared spontaneously in the 1920s in a flock in Florida.

Additional color varieties, in which the black is replaced with red, slate, and brown, have been developed. A variety that reverses the black and white has also been reported.

Just as Silver Narragansetts naturally occur in Narragansett flocks, gray Royal Palm sports occasionally occur in Royal Palm flocks. A sport is an unexpected color variation that appears spontaneously in an otherwise normal flock. Their similarity suggests a historic and genetic relationship. The Royal Palm was admitted to the *Standard* in 1977.

Nebraskan or **Nebraska Royale** turkeys are black-and-white spotted. Buff-and-white spotted turkeys are Spotted Nebraskas. Royal Nebraskas are charcoal gray or blue with white scattered through the plumage.

Black turkeys were popular in Europe, and colonists brought them to America in the seventeenth century. They were among the turkeys recognized in 1874 by the APA's *Standard of Excellence*. Poults may have some undesirable white feathers, but they usually molt to black, so give them time to mature before culling. *Shutterstock*

Blue Slate turkeys include birds that have blue feathers and those that have black specks on gray or lighter blue feathers. Hens are lighter than toms. *Shutterstock*

Black turkeys are also called Black Spanish and Norfolk Black, reflecting the locations where they were popular centuries ago. England, Spain, France, and Italy all favored Black turkeys in the sixteenth and seventeenth centuries. This is the turkey that colonists brought with them when they crossed the Atlantic. It was dominant in early American domestic flocks. Its feathers are glossy, iridescent black. No white feathers are permitted in exhibition today, although Norfolk Blacks of the past often had some white feathers. Its size is the same as the Narragansett.

Slate turkeys, also called Blue Slate, Blue, or Lavender, have light or ash blue plumage, in some cases flecked with black.

The color, or phenotype, of blue-colored birds results from two different genotypes, one dominant and one recessive. This is an important point for breeding flocks. Expert breeder Paula Johnson of New Mexico found that the ones with black specks, which she designates Slate, breed true, with all offspring being like their parents. When Blue turkeys with solid feather color are bred to each other, the offspring may be Black, Blue, or Slate. This variety, because of its variability, can be a challenging breeding project. Its striking and unusual color, however, can make it worth the effort.

They are sometimes called Gray turkeys, but that name is more correctly applied to birds with patterned feathers in shades of gray. The Oregon Gray has black-and-white feathers.

Lilac turkeys are slate or blue with light maroon or red barring.

Red, shading to Buff, dominates the coloration of several historically important turkey varieties. The **Bourbon Red** was developed in Pennsylvania and carried west

The Bourbon Red became a dominant commercial variety in the early twentieth century, before it was overwhelmed by the Broad Breasted Bronze. It has white primary flight feathers and white on its tail feathers. Underneath the rich mahogany outer feathers, the feathers next to the skin are light buff to white, making the carcass easy to dress out without dark pinfeathers. It is recovering in popularity and makes a good choice for small flocks. *Lin Ennis*

and south with settlers, eventually taking its name from Bourbon County, Kentucky. The variety was developed from selective breeding for darker color from the Buff turkey in the late nineteenth century. They were admitted to the *Standard* in 1909. The antecedents of Bourbon Reds were called Tuscorora and Tuscowara Reds, with breeders eventually settling on the Kentucky appellation. Regal Reds are similar to Bourbon Reds, from old varieties known as Arkansas Reds and Kentucky Reds, but their feathers are white next to the skin. The poults start out white and gradually molt to red. They sometimes appeared as sports in Bourbon Red flocks. Adult birds should not have white on the wings or tail, and breeders have been working to remove the white and purify the strain. Because the red color replaces the white over

time, these birds must be allowed to reach full maturity before they are bred. Dr. Tom T. Walker of Texas is convinced the Regal Red is the oldest red breed.

The Harvest Gold, a result of crossing a Regal Red with a Black Spanish, has brilliant gold in its plumage. It is a new color variety being raised by fanciers. "The extent that color and color combinations are hidden away in the Wild turkey and early Bronze variety is unbelievable," says Dr. Walker, the first breeder of Harvest Gold birds. "It's exciting to see how these colors reveal themselves in various matings."

Buff turkeys, also called Jersey Buff and New Jersey Buff, are solid buff-colored, although the wings may be lighter, even white. Their light feathers produce a clean carcass unmarked by dark feathers. Although this

The name Chocolate for this variety of turkey describes the color of its feathers, shanks, and feet. It was developed in the American South prior to the Civil War. The destruction caused by that war took the Chocolate turkey with it, and the breed remains very rare today. *Mike Walters*

Buff turkeys can be bred to a wide range of color intensity. Browner versions blend toward Chocolate. Original Buff turkeys, included in the first APA *Standard of Excellence*, were the progenitors of Bourbon Red turkeys, which replaced them in popularity. Renewed interest in Buff coloration brought back the Jersey Buff in the 1940s. Their light pinfeathers result in the clean carcass favored by consumers. *George McLaughlin*

variety was recognized in the first American *Standard of Perfection* in 1874, the Buff was not widely raised and became rare by the turn of the nineteenth century, eclipsed by the Bourbon Red. Many buff-color variations result from crossing red and black turkeys, usually with white or light-colored wings and tails. Darker buff-colored birds are called Fawn. A browner variety is the Clay or Claybank, which was common in the South in years past. Darker still is the Chocolate, which traces its history back to the antebellum South and across the Atlantic to France.

Wild turkeys tame readily, especially when raised from poults. Tales proliferate of wild turkeys joining domestic flocks for the mating season or longer. Other wild poultry breeds, such as Mallard ducks and Canada geese, have been recognized for production and exhibition, but wild turkeys have not been. Pennsylvania State University developed a domesticated wild turkey in the 1930s called the Nittany turkey.

Wild turkeys, compared to domesticated varieties, may be more reactive to sights and sounds around them, but they could still be a good choice for small flocks.

Some flocks listed as wild today likely descend from Pennsylvania's Nittany.

Geographic separation has resulted in several different color varieties of wild turkeys. Wild turkeys are smaller than domestic ones, although very large examples are recorded. Generally, around 20 pounds is the largest they grow. The **Eastern** wild turkey ranges along the eastern half of the United States and Canada. Its tail coverts, the smaller feathers covering the bases of the main feathers, are chestnut brown, tipped with velvet black. Both its primaries and secondaries have white barring. It can stand as high as four feet tall. The **Osceola** or **Florida** turkey is smaller, with green-and-purple iridescent feathers that have less white and more irregular barring. Its tail coverts are also chestnut. It is found only on the Florida peninsula. The **Rio Grande** turkey, native to the central plains and ranging to elevations of six thousand feet, is distinctive because of its long legs. "You have only to see them race across an open space like a racehorse breaking out

All turkeys are the same breed, so this list identifies varieties that have historic value or are rare. All traditional turkeys are rare. Bourbon Red, Bronze, and Royal Palm are the best represented of the group. Turkeys have received breeding attention from commercial interests, which have developed new varieties, such as Beltsville and Oregon Gray. Other varieties have resulted from small-flock breeders, such as Apricot, Chestnut Blue, Harvest Gold, and Sweetgrass. They are rare but not old, having been developed in the last century.

Grizzled and Spotted turkeys have long histories and are old. The name *Nebraska* was added in the twentieth century. Grizzled turkeys are sometimes called Royal Nebraskas, and Spotted turkeys are sometimes called Nebraska Royals. Go figure.

Apricot	Rare
Auburn	Old and Rare
Silver Auburn	Old and Rare
Beltsville (White Midget)	Rare
Black (Black Spanish)	Old and Rare
Black Wing Bronze (Crimson Dawn)	Old and Rare
Bourbon Red	Old and Rare
Bronze	Old and Rare
Blue Bronze	Old and Rare
Buff	Old and Rare
Calico	Old and Rare
Chestnut Blue	Rare
Chocolate	Old and Rare
Harvest Gold	Rare
Lavender	Old and Rare
Lilac	Old and Rare
Mottled	Old and Rare
Narragansett	Old and Rare
Silver Narragansett	Old and Rare
Grizzled Nebraska	Old and Rare
Spotted Nebraska	Old and Rare
Nittany	Rare
Oregon Gray	Rare
Regal Red	Old and Rare
Royal Palm	Old and Rare
Blue Palm	Old and Rare
Red Palm	Old and Rare
Slate	Old and Rare
Sweetgrass	Old and Rare
White Holland	Old and Rare

At least six subspecies of wild turkeys are identified in North America and Mexico. This Eastern wild turkey was significant to Native Americans, who used its feathers for decoration and religious purposes. Turkeys are symbolic of the earth's bounty and the blessings of the harvest. They can be profitable for small-flock farmers. *Shutterstock*

of the chute to realize that the Rio Grande has a different structure from other wild turkeys," says Dr. Walker. Their green feathers have a coppery gleam in the sunlight, shading to light buff at the tips of the tail and lower back. **Merriam's** turkey inhabits the Rocky Mountains and high plains. It has purple-and-bronze feathers tipped in white on the tail and lower back. **Gould's** turkey lives in the Southwest desert of the United States and Mexico. It is larger than its wild relatives, with long legs and large feet. Its plumage is copper and greenish gold, with long tail feathers. The upper tail coverts are tipped in almost white pink. Its outer secondaries have broad white edges.

The **South Mexican** turkey is native to Mexico, where it was the original bird domesticated by the Aztecs. It has less red in its black tail feathers than other wild varieties. White appears only in the margins of its secondaries.

FEEDING

Turkey poults grow fast, so they need high-protein feed to keep up. As they grow, their needs taper off. They do better on game-bird feed, with its higher protein content (up to 28 percent), than chicken feed. They can be tapered off it after eight weeks to grower crumble or pellets with 22 to 24 percent protein. The protein content should taper down to 18 to 20 percent around twelve weeks of age and after sixteen weeks to layer crumble, scratch feed, and pasture. Adult birds need 15 to 17 percent protein. Commercial preparations are good and give precise directions.

If turkeys are on pasture and not crowded, they will get some protein from the insects and worms they forage. Breeder flocks can be left on pasture during the winter. Winter wheat grown out at least 6 inches tall, alfalfa, and other grains make good forage for them.

Pastures should be rotated to avoid overgrazing and parasite infestation.

BREEDING

Heirloom breeds will mate and brood naturally. Keeping a historic variety makes you part of livestock conservation. Each additional flock increases the genetic strength of these birds.

Although commercial Broad Breasted White turkeys are always artificially inseminated, traditional breeds can manage mating on their own. One tom can mate as many as ten hens, but usually two to four are enough.

Hens need about twelve weeks of rest during the winter and do not lay during the short days. They molt during this time and prepare for the breeding season. They begin to lay in May. Artificial light can extend the day length to as much as fourteen hours, providing the photostimulation toms need for adequate sperm production and for hens to lay eggs. Toms respond more slowly to extended days and may require eight weeks of photostimulation to produce adequate sperm.

A tom turkey's head shows the snood hanging over the beak. The wattle is the fleshy part dangling down from his neck. A turkey's head changes color in response to aggressive challenges. Some have blue pigment, which can be seen on this bird, and their heads turn bright turquoise blue. Others become bright red. *Shutterstock*

Roosts

Turkeys are perching birds that naturally roost in trees. Poults as young as two weeks old will look for a roost. They can be accommodated with 2-inch-diameter poles or branches set 6 inches off the ground. Allow 6 inches per bird. Mature birds need stronger roosts that will support them. Flat frames of 2-inch poles or two-by-fours 24 inches apart, 15 to 30 inches off the ground, will suit them.

Turkey hens are often good brooders and good mothers, like this Royal Palm hen. Eggs incubate for twenty-eight days, much like chicken eggs. Poults are ready to go outside and learn to forage right away, but they are fragile and subject to weather. Make a secure house available to them and lock them up at night. *Shutterstock*

Small-flock owners find it convenient to rely on the sun in spring and summer and extend the laying season into the fall with artificial light if that suits their plans. Compact fluorescent lightbulbs that use less energy are available from poultry supply houses. One 50-watt bulb per 100 square feet should produce enough light.

Turkey eggs can be hatched naturally by turkey hens, by broody chickens, or in incubators. Turkey poults are fragile and need protection for the first two months. A mother will keep them warm and protect them, provided she herself has adequate feed, water, and shelter. Alternatively, poults can be purchased from hatcheries. Brooders must maintain a constant temperature, preferably thermostatically controlled. The poults need this additional warmth until they are four to five weeks old.

Poults can be raised on deep litter or frequently changed absorbent litter. Avoiding contamination from

droppings is essential. Wire mesh can effectively keep poults away from soiled litter. Their tiny legs require ¼-inch mesh for the first three weeks, then ½-inch mesh until they can go outside. Avoid slick flooring, such as newspaper, on which their feet slip. Their legs can be permanently damaged from slippery flooring, resulting in a condition called splay leg.

HEALTH MANAGEMENT

Turkeys do not require routine vaccinations. However, vaccines are available for several common diseases, including fowl cholera, turkey pox, avian encephalomyelitis, and Newcastle disease. Check with local extension agents and veterinarians to determine whether such protection is warranted in your area.

Blackhead is the colloquial name for histomoniasis, a serious protozoan parasitical disease that has been

Watering

Hanging feeders and waterers, adjusted to the height of the birds' eyes as they grow, will reduce the amount of feed and water wasted as the birds dig around in it with their beaks. Sloppy waterers leave wet litter to ferment and foster disease-causing organisms. Feeders on the ground should not be filled more than half-full, to keep feed contained.

Turkeys in the Rain

Are turkeys so stupid they will turn their heads upward and drown in the rain? Have a heart attack if an airplane flies over them? "These old 'farmer tales' are so ridiculous, yet I hear them all the time," says poultry expert and 4-H leader Jeremy Trost of Wisconsin. "Think about the survival of the fittest. If a turkey really was so dumb as to drown in the rain, would they be on the earth today?"

Dr. Tom Walker, now in his eighties, remembers an incident from when he was a teenager. His flock of turkeys ranged far from the safety of home, foraging all day on delicious bugs and grasshoppers. They reliably returned home at night to enjoy the extra feed he gave them. When a sudden Arkansas thunderstorm opened the heavens on the farm one day, he feared that the young poults would turn their heads up and drown. With more poults than the hens could cover in the flock, he ran out into the rain to save them.

He found them, huddled under a tree, with the old tom sheltering all the poults under his fully spread wings.

"The mother hens were standing by helpless because the young turkeys had chosen the old tom's umbrella," he says. "In the years before and all the years since, I have never seen that at any other time."

One consideration that may have encouraged this myth is the fact that young turkey poults do not develop sufficient oil on their feathers until they are around eight weeks old. If left outside in wet weather, their feathers will get soaked, and the birds may get chilled and die.

As with all animals, appropriate care is necessary. Poults' immune systems need to mature, which happens around eight to twelve weeks of age. Those being raised by mother turkeys or foster chickens can and should go outdoors a day or two after hatching for short periods. Their experiences allow them to develop natural immunity to diseases they will face in that environment. However, they are fragile in their early weeks and need protection from cold, heat, rain, and germs.

Keep them clean and dry. Feed them well. Don't worry about them drowning in the rain.

Turkeys like to have room to roam. Fence them with netting that is small enough that they can't get through. Electric fences can help keep turkeys in and predators out. Natural barriers like hedgerows can be part of the boundaries, but determined turkeys can work their way through very dense brush.

The amount of range needed depends on the characteristics of your property. Consider the well-being of the turkeys and what you will need to manage them effectively. The fundamentals are to get them food and water and keep them safe.

Turkey pasture generally should be limited to 100 birds per acre, although lush pasture may be able to support 125. Less verdant fields may support only 50 birds. Pasture should be allowed to rest during the winter, after the birds have gone to market. Turkey pasture can be seeded with alfalfa, clover, orchard grass, or other locally successful native grasses. Confer with local experts to determine what is best in your area. More range allows the birds to forage for more of their own food.

Avoid damp locations. Damp or wet areas breed turkey pathogens and pests. Turkeys need shade to get out of the sun on hot days. Their natural habitat is brushy cover.

Raising turkeys in confinement requires 5 square feet per bird for toms, 3 square feet for hens.

Shutterstock

afflicting turkeys as long as records have been kept. It's the main reason turkeys should not be kept with chickens. Chickens can be hosts to the cecal worm, the eggs of which can harbor the parasite. The protozoans multiply and infest the liver and intestines. Symptoms include drooping head and brownish or bloody, foamy droppings, but the first thing you notice may be dead birds. The head may turn dark, but that does not happen in all cases.

These microscopic organisms can persist in the soil for years, so turkeys must be kept off fields that have any history of blackhead. Earthworms can also carry the eggs, engendering contamination. Prevention is best, by worming the flock to reduce the cecal worms that carry the parasite. Avoid manure buildup. If birds are confined, keep them on three inches or more of crushed limestone. Keeping birds on wire also helps them avoid contact with droppings.

Mycoplasma-related diseases are caused by extremely small organisms that can infect the flock and retard growth, reduce egg production, and make the birds sick or even kill them. This class of pathogen can cause sinusitis, also known as chronic respiratory disease, and infectious sinovitis, which causes lameness, among other diseases. If the flock gets infected, they will have to be killed and your facilities disinfected before starting over. It's spread in dust and dirt that can be tracked around on people's feet and clothes, trucks, or anything that comes to your farm. Avoid it with basic biosecurity practices.

Bluecomb and **mud fever** are colloquial names for turkey coronavirus, a contagious diarrhea that can kill poults. Adults usually survive, but chronic flock infection causes poor growth. If infected, the flock may have to be killed and the facilities disinfected and allowed to dry for four to six weeks before starting over.

HOUSING

Turkeys need protection from the elements and from predators. Those living on range will learn to come to the shelter when they consistently find feed and water there. They should be secure in their shelter an hour before sundown.

Movable shelters like chicken tractors can house turkeys as effectively as they do chickens. Turkeys do not adapt readily to change, so the shelter should be moved gradually, a small distance each day. If moved a longer distance, the turkeys may get confused and have trouble locating it. In the event it needs to be moved a long distance quickly, confine the turkeys in it for a few days so that they become accustomed to the new location. Allow no less than 3 square feet per bird.

TURKEYS AND PEOPLE

Turkeys are friendly and personable. They make good interpretive birds and, with attention, enjoy visiting schools and other public places. At Garfield Farm in La Fox, Illinois, visitors enjoy seeing Narragansett turkeys as part of the farm's historically accurate breeding program. At Claude Moore Colonial Farm in McLean, Virginia, Bronzeback (an old name for Bronze) and Spanish Black turkeys are part of the interpretive program. "They work very well as loose livestock around the public because they are fairly calm and naturally curious," says J. D. Engle, facilities manager. "They find the visitors almost as fascinating

as the visitors find them, and they'll just walk away if someone gets too close for their comfort.

"We train the turkeys from the time they're poults to herd and stay with people," says Mr. Engle. Children are allowed to herd them to the tobacco fields to forage on the hornworms that would otherwise devastate the crop. "I have seen a hen leap five feet straight into the air to pluck a worm off the top of a tobacco plant," he says.

Herbert Myrick recognized the problem of maintaining standard breeds a century ago in *Turkeys and How to Grow Them*: "When there are no wild turkeys except for those preserved by man, the salvation of the domestic turkey will depend on fanciers—those who breed for beauty and utility. They maintain the varieties pure and perfect them. They, only, expend the required time and money, and follow the laws of breeding necessary to prevent the stock from running out. When will farmers, generally, appreciate the value of such service and cease to scoff at fancy prices?"

These Spanish Black turkeys at the Claude Moore Colonial Farm in Virginia are picking destructive worms and other insects from the tobacco plants in the farm's field. They are historically accurate for the farm's eighteenth-century period. Turkeys are effective pest control birds, although they can also be destructive to plants. Consider your own operation and how turkeys can be a working part of the program.
Jan Tilley, Claude Moore Colonial Farm

CHAPTER 7

GUINEA FOWL

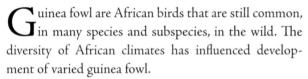

This Black-Legged Tick, also known as a Deer Tick, is the one primarily responsible for transmitting Lyme disease. It is very tiny, comparable in size to a grain of black pepper. Like other ticks, they set on the edge of a leaf and attach themselves to a potential host as one passes by. Unless a guinea gets them first! *Scott Bauer, USDA Agricultural Research Service*

Guinea fowl are African birds that are still common, in many species and subspecies, in the wild. The diversity of African climates has influenced development of varied guinea fowl.

Many people keep them as insect-control birds. They eat all kinds of pests, including deer ticks, an important point for those who live in areas threatened by Lyme disease. They happily consume Japanese beetles, wasps, and other pests but, unlike chickens, don't scratch and dig up the garden.

Guinea fowl are credited with killing and eating small snakes and rodents. Sharon Wilson of Texas witnessed them killing a six-foot snake at a guinea fowl farm. Their screeching alone is said to discourage rodents. Jeannette S. Ferguson has written an entire book called *Gardening with Guineas*.

Then again, their screeching may discourage you and annoy your neighbors. My husband finds them unbearably annoying. R. H. Hastings describes their constant chatter as "A running commentary on the nature of the food" that "will fetch other guineas from many yards away to share the delicacy." The positive side of this characteristic is that they serve as excellent watchdogs and will warn you of any unusual occurrence on your farm.

Guinea fowl like their own reflections, so if they are settling in places you don't want them, try hanging a mirror where you want them to nest. They will find it and move.

HISTORY AND CULTURE

Fossilized remains of guinea fowl date back 2.5 million years, in what is now the Czech Republic, where they roamed the landscape with prehistoric elephants and lions. Indian and Burmese people may have kept guinea fowl since Neolithic times, as early as 7000 BC. They were domesticated at least four thousand years ago, showing up in an Egyptian pyramid mural of 2400 BC. The nobles of that period, the Fifth Dynasty, enjoyed maintaining aviaries. Some guinea fowl were indigenous to the area, and others were imported from further south in Nubia, Sudan, Ethiopia, and Somalia.

With the arrival of early domesticated chickens by 1475 BC, poultry husbandry increased. Egyptian incubators of the time could accommodate up to ninety thousand eggs, both guinea fowl and chicken eggs together. Guinea fowl even appeared in hieroglyphics.

Whether Greek farmers acquired guinea fowl from Egypt or elsewhere, the birds were being raised on Greek farms by 400 BC. Guinea fowl are among the birds

The Pearl guinea fowl is the common domesticated guinea fowl most often found in small flocks. It is a good table bird for home use, but it has fallen off popular consumer awareness. Increased interest in local food and a willingness to try different kinds of poultry may help restore guinea fowl to the popularity it enjoyed over centuries of domestication. *Shutterstock*

associated with the Greek goddess Artemis. In one tale, the sisters of the hero Meleager wept themselves to death after Meleager died. Artemis rescued them from Hades and turned them into guinea fowl. The spots on their feathers represent the sisters' tears. Whether this myth accounts for the scientific name for Helmeted guinea fowl, *Numida meleagris*, or the name is a corruption of melanargis, meaning black and white, remains unresolved. Domestic guinea fowl were developed from the Helmeted species.

Roman writers like Horace, 23 BC, and Pliny in his natural history, AD 77, mention eating both the meat and eggs of guinea fowl. Romans likely distributed the birds across Europe with the spread of their empire. As the Middle Ages advanced, guinea fowl disappeared from the menu. Portuguese traders reintroduced guinea fowl in the fourteenth and fifteenth centuries. Turkeys were arriving from the New World at the same time, resulting in some confusion as to what they were. That confusion was reflected in their names, with turkeys eventually being designated *Meleagris gallipavo. Pavo* is the Latin and Spanish word for peacock, so the Spaniards who brought turkeys back to Europe initially called them *pavo de las Indies*, the "peacock of the Indies." The popular name *turkey* already meant guinea fowl in Europe. The American bird soon won that name, but the confusion lingered.

Turkeys and guinea fowl aren't related, except in the sense of both being fowl. They both just looked weird to Europeans.

Guinea fowl traveled to America with the slave trade, where feral flocks were established in the Caribbean islands during the 1700s. In China, they became known as Pearl Fowl and were quickly adopted. By the eighteenth century, they were widely cultivated.

Victorian England raised the guinea fowl to the status of one of its most popular table birds, a luxury item with prices at record highs. Production has declined since then, but interest in gourmet, alternative, and local food could open new markets. Modern breeding practices and selection have developed guinea fowl varieties that mature more quickly and extend the laying season to provide eggs and chicks for production almost year-round.

Despite centuries of domestication, guinea fowl retain a lot of wildness. They prefer to range free, and they fly well. Most owners allow them liberty and occasionally find them roosting on the barn or house roof and in trees. Long-time SPPA member Donnis Headley describes the experience of having guineas as comparable to "an eccentric but wealthy relative who has come for an extended visit—tolerated, even welcomed, in the unspoken hope that one day their benefits will outweigh their inconvenient behaviors."

Guinea fowl are most at home ranging free. In their native East Africa, the springtime emergence of flocks of guinea fowl into the fields from the woodlands was the sign that it was time to begin planting. Natives considered that emergence, usually occurring at the beginning of February in the developed world, to be the beginning of the year.
Shutterstock

RAISING KEETS

When raising guinea fowl, it's best to start with babies, which are called keets. That gives you the opportunity to tame them and give them the basics of training.

You can purchase keets from a breeder or hatch eggs. Chickens make good foster mothers, if you are already established with chickens. A large-breed chicken may be able to cover as many as forty guinea eggs. Chicken eggs have an incubation period of twenty-one days, whereas guinea eggs require twenty-six to twenty-eight days to hatch, but a broody hen will not mind the week longer. Artificial incubators are also successful. Guineas also like to nest and raise their own keets. Guinea hens can be very secretive about making nests for themselves, but unfortunately, they may not choose locations safe from predators. Jeannette Ferguson recounts in her book *Gardening with Guineas* the experience of one guinea who succeeded by building a nest under a pile of barbed wire. That's the exception to the rule, however. When a hen doesn't come back to the house at night, follow the rooster who is guarding her. Guineas are mostly monogamous, and he will be protecting the hen on the nest.

The eggs are smaller than chicken eggs, weighing 1.4 ounces to the 2-ounce large chicken egg. They have thick shells that make them difficult to candle. Use a bright light source on the tenth day of incubation to check for development.

The keets are very tiny and must be managed carefully. A foster chicken mother may do a better job than a guinea mother and father. Domestic guinea fowl are often not very good parents, although both participate in raising the keets. Because their origins are on dry grasslands, they don't manage moist conditions like dewy wet grass well. Keets may die from getting wet and chilled. A guinea mother may leave the nest before all the keets are hatched. Confining the family in a pen on bare ground or short grass until the keets are fully feathered, around six to eight weeks old, improves the parents' chances of raising their family successfully.

Keets should be kept on cloth or plastic with a rough texture so that they can get traction without falling through. Their legs are weaker than chickens', and early leg injuries will persist throughout life.

Guinea fowl like this Pearl couple make good parents, provided they get the support they need. As dry desert birds, they lack understanding of wet conditions, which can be fatal to keets. Keep them enclosed on sand or dirt until the keets are fully feathered. Keets vocalize constantly, staying in touch with their mothers and allowing her to exercise her vigilance over them. Color varieties differ from the start, but similar colors are difficult to distinguish from each other until the keet molts to adult plumage at two to three months of age. *Shutterstock*

Training

Guinea fowl have their own view of the world and their place in it. They retain an independent nature, although training makes them more manageable. Jeannette Ferguson recommends white millet as a special treat to give some measure of influence over guineas. Associate yourself with food treats, and they will learn to come when called.

Guinea fowl do best with people if they are raised with a lot of gentle handling. Spend time with them and handle them daily. Otherwise, they may be so skittish you will have difficulty approaching them.

FEEDING

Keets do well on turkey starter or game-bird feed, which have 24 to 26 percent protein, until they are fully feathered and can be set outdoors. Medicated feeds are not recommended, since they are not formulated with keets in mind, and the birds may ingest an overdose when they go through growth spurts and eat more.

On sufficient range, keets will eat enough bugs and seeds to satisfy the nutrition needs of their entire diet. If they don't have enough to eat, supplement them with chicken layer crumbles or game-bird feed with higher protein content. They need grass. If they are confined or grass is limited, give them alfalfa hay.

They are willing consumers of all kinds of kitchen trimmings. Avoid giving them anything that you don't want them to help themselves to later in the garden. They like fruit and berries but don't usually bother vegetables much. Chopped garlic and onions reduce worms and may help birds resist respiratory problems and coccidiosis. Edible seaweeds, such as kelp, are welcome sources of minerals.

Keets will also eat bees, so keep hives and guineas separate if you are raising honeybees at the same time.

HOUSING

Guineas raised for meat and egg production can be confined as other poultry are. Small-flock owners usually let them range. Their love of tasty food can be used to train them to come when called. Although they prefer to roam free, they are vulnerable to predators and should be secured in a protective shelter at night and during harsh weather.

Greens Favored by Guinea Fowl

Shutterstock

Dandelion
Land cress (American cress)
Watercress
Spurrey
Cleavers (goosegrass)
Chickweed
Couchgrass
Fat hen (lamb's quarters)
Fennel
Wormwood
Vetches
Red clover
Fenugreek
Thistleheads and seeds
Dill
Wild fruits, including rose hips, blackberries, bilberry, and cranberry families
Licorice
Dried edible fungi
Iceland moss

Herb grasses
Sorghum
Brown rice
Corn and all farm grains
Palm kernels, palmyra
Soybeans
Seeds of all kinds: sesame, cow peas

Guinea fowl are good flyers, although they are more inclined to run away from predators than fly from them. They are busybodies and like to look over the farmyard from a high perch. This Pearl Gray guinea fowl is one of the most popular varieties and one of the first recognized by the APA for exhibition.
Robert Gibson

Although they fly well, they generally try to escape from predators by running, often unsuccessfully. Left alone, they will roost at night in trees, where they are vulnerable to owls and hawks.

One Texas breeder uses an old-fashioned buggy whip to herd his birds into their house for the night. He scatters grain on the ground and then uses the whip as an extension of his arm to guide them toward the house.

An existing outbuilding can be converted to a guinea house, or you can build one for the purpose. Confined birds should have 3 to 4 square feet per bird.

Surround the house with a fenced yard. Allow 20 square feet per bird in a breeding pen. Natural shrubs and trees inside the enclosure provide guinea fowl with places to perch. Shrubbery and climbing plants planted around the outside can be trained to grow over the fence. A sandy spot in the sun for dust baths helps keep their feathers in good shape. Guineas will happily fly over the fence unless the aviary is completely enclosed. If you allow them free range, accommodate them with two-by-fours around the top of the fence, to land on as they make their way over.

Guineas prefer to roost as high as they can. "All species of guinea fowl are always eager to see what is going on around them," writes R. H. Hastings Belshaw in *Guinea Fowl of the World.* Provide at least 10 inches of roost per bird.

No Place Like Home

Adult guinea fowl resist learning a new home and may continue to leave a new site in search of their former home at every opportunity. You might become a guinea owner when you buy a farm that has resident guineas. They come with the property. However, new birds need six weeks confined to their house and yard to confirm to them that this is their home. This gives them the opportunity to commune with other birds and learn their new yardmates.

The Pied color varieties combine patches of white with the solid colors, such as the unusual expression of crossing on this bird. Separate genes influence the white patches, the spots, and the background color. A separate gene for tone changes the intensity of the color expression. For example, Bronze and Purple guinea fowl have the same color gene, tempered by the modifier gene that affects tone. As a result, many guinea fowl plumage variations are possible.
Shutterstock

BREEDS AND VARIETIES

There are six separate species of guinea fowl: White-breasted guinea fowl, *Agelastes meleagrides*; Black guinea fowl, *Agelastes niger*; Helmeted guinea fowl, *Numida meleagris*; Plumed guinea fowl, *Guttera plumifera*; Crested guinea fowl, *Guttera pucherani*; and Vulturine guinea fowl, *Acryllium vulturinum*. The Helmeted guinea fowl gave rise to today's domestic varieties and is the species most commonly kept, but fanciers also keep Crested and Vulterine guinea fowl. The others are rarely kept in captivity outside of zoos and game parks.

Guinea fowl are raised in many colors: Pearl Gray, White, Lavender, Royal Purple, Coral Blue, Buff Dun-dotte, Buff, Porcelain, Opaline, Slate, Brown, Powder Blue, Chocolate, Violet, Bronze, Sky Blue, Pewter, Lite Lavender, and Pied. Pearling refers to the white dots on the feathers. Pied birds have white patches in otherwise colored plumage.

Guineas are exhibited at poultry shows. The *Standard of Perfection* recognizes three colors for exhibition: Pearl, Lavender, and White. Weight is a breed characteristic,

and exhibition weights range from 3 to 4 pounds. Underweight and overweight birds are penalized.

Males and females are very similar in appearance. Males develop larger wattles, but that's a relative quality. Only females make the shrill "come back, come back" call. Some start making this call as early as six weeks old, but others may not start calling until they are older. The male has only a single call, a high-pitched single syllable. Hens can successfully imitate it, so rely on both larger wattles and sound to determine which you have.

Solid-color Slate birds are a domestic variety of guinea fowl. They lack the spots found on the more common Pearl variety, which is the wild coloration. This bird belongs to Tomas Condon, a University of Connecticut student who is making rare fowl his life work. *Tomas Condon*

This Pied Pearl guinea is also known as a Silver Wing or White Breasted because of the location of the white patches. Breasts and flight feathers are the most usual location for white patches. In France, white-breasted birds are called "Buonaparte." *Shutterstock*

The pure White guinea fowl is a historic favorite. It is a natural color mutation that achieved popularity in the early nineteenth century in Europe and Africa. Crossing it with Pearl and other varieties results in white-fronted offspring. *Shutterstock*

The male's call is different from the female's, which helps distinguish them from each other. The hen has a two-syllable call, variously described as "come-back" or "buck-wheat." Both males and females have the single syllable machine-gun "chi-chi-chi" call. *Shutterstock*

PRODUCTS

Guinea fowl are in demand as table birds at gourmet restaurants and retail markets, if you are able to make the connections and develop the niche. They are seasonal egg layers, naturally laying from the end of March through mid-May, with the potential of producing as many as one hundred eggs a year.

Despite being limited by the season, the unusual dark shells and small size of guinea eggs make them eye-catching for specialty retail and restaurant trade. For cooking purposes, two guinea eggs are equivalent to one chicken egg.

Guinea hens, left to their own devices, lay about thirty eggs at the rate of one a day and then go broody. To sell eggs for food, gather them at least once a day. Leave four to six dummy eggs in the nest to attract the hen to return and lay in that nest. Egg-layers may be kept confined each day until afternoon to persuade them to lay in the house or yard. Continually removing eggs extends the laying season, as long as to October.

As meat, guineas raised on broiler diets are large enough to be processed as early as ten weeks of age. Breast meat increases if raised to sixteen to eighteen weeks of age. Smaller birds may be marketed as substitutes for partridge or quail. Live weights of 2¾ to 3¼ pounds produce dressed weights of 2¼ to 2¾ pounds, suitable for a meal for four people. Currently, dressed birds weigh 3 to 4 pounds and sell at retail for more than $8 a pound.

The meat is similar to that of other game birds in flavor, according to *Worldwide Gourmet*. Nutritionally, it is lean and low in sodium. Depending on the guinea's diet, the meat may also be high in fatty acids. It is low in calories at 134 calories per 100 grams, or 3.5 ounces. Comparatively, turkey has 109 calories per 100 grams.

The feathers of guinea fowl are sought both by flower shops and craft and specialty businesses. Lavender, Purple, and Blue colors, and Pearl color varieties with their attractive dots, are especially in demand.

Lavender guinea fowl plumage is a pale silvery gray with spots. The skin is almost white, paler than the usual pink, with a blue tint on the breast, limbs, and back. *Shutterstock*

Guinea fowl can be raised in confinement as broilers for the gourmet market. Their bones are small and the meat is flavorful, somewhat gamy. Lighter-colored varieties have lighter-colored meat. *Shutterstock*

CHAPTER 8

• • • • • • • • • • • • • • • • • • •

GAME BIRDS

The common quail is native to Asia but is a happy import to the United States, where it is hunted for sport. Such activities have a long history in the relationship between humans and birds. *Library of Congress*

Game bird is an informal name that refers to any wild bird that has been hunted for food, sport, or feathers. In the world of raising poultry, it includes peacocks, pheasants, quail, partridge, and pigeons. Except for pigeons, which are passerines, they are all gallinaceous birds, the relatives of domestic fowl. Many tame readily but remain wild birds. A more technical term for this group of birds, but less commonly used by the farmer, is *land fowl*, which distinguishes them from waterfowl and ratites.

Game birds are available in many colorful species and breeds. In culinary matters, some species are farmed for the table. Strains being raised for those purposes are selectively bred for carcass size and meat. Peacocks and pheasants are bred for their feathers, although many are raised solely for their beauty and spiritual value.

Peacocks, pheasants, and partridge are not native to North America but have successfully established themselves as residents. Many are native to Eurasia, ranging from the Black Sea across the Himalayas, throughout India and China to Japan. Others are native to Southeast Asia and the island nations of Sumatra, Borneo, Java, and Indonesia. The Chukar partridge is the national bird of Pakistan.

The natural habitat of land fowl ranges from jungle to grasslands to shrubs, but they are primarily ground-dwelling birds that prefer to stay in the underbrush. Like chickens, they enjoy taking dust baths. They nest on the ground but at night roost in trees or other locations above their natural predators. Mothers stay on the ground with their chicks as long as necessary. The young are usually able to fly within ten days.

Being gregarious in varying degrees, all have many vocalizations to communicate with each other. William Beebe, in *A Monograph on Pheasants*, observes that many pheasants continue their vocalizing even when they are alone. In the wild, they rely on sight and hearing to alert them to lurking predators. Captive-raised birds remain alert to environmental cues of sight and sound. They fly up into trees to escape danger.

PEAFOWL

Peacocks are the males, peahens are the females, and peachicks are the offspring of peafowl. They are considered members of the pheasant family, although molecular

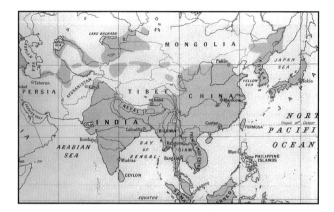

This map shows the distribution of the Phasianinae, from William Beebe's *Monograph of the Pheasants*. The pheasant family includes pheasants, peafowl, junglefowl, partridges, and quail. Many of them tame easily, but only a few truly domesticate and become commercially significant. Others are kept by professional institutions and master aviculturalists for conservation purposes. As with other wildlife, their existence is threatened by loss of habitat and environmental degradation. *William Beebe,* Monograph of the Pheasants

The Blue peacock is the most frequently kept member of the peafowl family. Some keep them as attractive pets living wild on their property. They become feral easily and need to be kept penned for several months before allowed freedom; unless they recognize their surroundings as home, they won't stay home. *Shutterstock*

research is providing new information as to the relationships between them and pheasants, jungle fowl, francolins, grouse, and turkeys. Adult Blue peahens weigh around 7.5 pounds and peacocks can weigh anywhere from 11 pounds to as much as 15 pounds. Their tails extend longer than 6 feet, and they have a wingspan of 4 feet. Peafowl are excellent flyers and will roost up high. Keep them penned for the first several months you have them, or they will become feral. Peacocks retain many wild characteristics and, if given the chance, will hide their nests and spend time roaming the countryside.

There are two Asian species of peafowl with many color variations. The India peafowl is the familiar type that has the turquoise blue neck feathers. Green peafowl are bright emerald green. All males have the elegant tail feathers, known as elongated tail coverts, which they can shake out and display. The eye spots on the tail are called ocelli.

The Congo peafowl, *Afropavo congensis*, is a wild species native to Africa. It was identified in 1936 but has not been kept as a domestic bird. With characteristics of both guinea fowl and peafowl, it may be a link between the two families. The International Union for Conservation of Nature (IUCN) lists it as threatened due to habitat loss, small population size, and hunting in some areas.

Indian peafowl, *Pavo cristatus*, occur in natural color mutations, eleven of which breed true and are considered varieties: India Blue, Oaten, Cameo, Purple, Buford Bronze, Peach, Opal, Midnight, Jade, White, and Silver Pied. The White is not an albino, since it has coloration in its eyes, around the ocelli, and on other parts of its body. It is a recessive variation.

Within each color variety, different patterns are possible, including Black-Shouldered, Pied, and White-Eyed. The Black-Shouldered pattern has solid wings rather

Peacocks have been bred into many different colors. The United Peafowl Association estimates as many as 185 color variations are possible. White is a natural mutation that prevents pigmentation in the feathers. *Shutterstock*

than lighter-colored barring. Pied means the bird has white patches in its plumage. The Silver Pied variety is almost entirely white, with some dark feathers in the tail. The eye in White-Eyed refers to the ocelli, not the bird's actual eyes.

The Green peafowl, *Pavo muticus*, is the subject of continuing discussion by those who have studied it. At least three distinct subspecies exist: the Javanese, the Indochinese, and the Burmese. Some authorities identify six separate subspecies. Green peafowl are listed as vulnerable by the IUCN and the Convention on International Trade in Endangered Species of Wild Fauna and Flora (CITES).

Most Green peafowl variations have a tufted crest, in contrast to the fanned crest of the Indian peafowl, and a distinctive yellow crescent or "war-stripe" on each side of the double-striped head, called the loral axe. Male and female Green peafowl are similar in plumage, except for the male's longer train. They have black wings with a blue sheen and pale brownish-yellow primaries. Males are substantially larger than females: as much as 10 feet long, including train, and weigh 12 pounds, compared to the female's 3.5 feet and 2.5 pounds. They are capable of sustained flight and in the wild are reported to fly as far as fifteen miles from island to island.

The India peafowl and the Green peafowl are separate species, and their natural habitats do not overlap. However, they readily crossbreed in captivity and among escaped feral birds. The resulting offspring led to the Spalding (or Spaulding) variation. Opinions are divided on Spaldings, although most advocates welcome both pure and mixed lines.

Conservationists are concerned that this interbreeding may contaminate the gene pool of captive birds and further confuse the complicated lines and disputes over species and subspecies. Fanciers revere the Spaldings for the improvement they bring by combining traits of both species. The wild, flighty, and delicate temperament of the Green is tempered by crossing with the India Blue, while preserving the astonishingly beautiful colors. Many breeders maintain both pure bloodlines and Spaldings. One breeder has imported Green peafowl from Germany having bloodlines that can be traced back hundreds of years. Chicks from those birds will be available in the future. A studbook is being kept by the United Peafowl Association.

"To fanciers, Spaldings are a very desirable bird," said Carol Cook of Cook's Peacock Emporium in Shorter, Alabama. "It takes several generations, ten years of breeding, to get a high-quality Spalding in a new color like Buford Bronze."

Peahens can raise their own chicks, or turkey hens make good foster mothers. The eggs have the same duration of incubation and turkeys and peafowl are susceptible to similar diseases. The inverse is not necessarily true (i.e. peahens cannot raise turkey chicks). Peachicks are more precocial than turkey poults, with wing feathers already in place at hatching. They require less solicitude from their mothers than turkey poults. *Shutterstock*

Because recessive genes may not be expressed in the physical characteristics of an individual bird, check the breeder's references if genetic purity is important to you. Mrs. Cook recommends prospective peafowl owners start with India Blues to learn the nuances of peafowl plumage and behavior. At around $100 a pair, or $150 for a trio, this variety is affordable. Exotic colors can cost hundreds of dollars more.

"If this is your first peacock, you need to be able to enjoy the bird and not worry about how much money you are spending," she says.

Generally, peacocks mature by three years of age, although some mature by two. Peahens hatched early in the season may lay fertile eggs at two years of age but will likely not lay a full clutch. The feathers that grow out of the male's back and are supported by his tail are called his train. The train is important to successful breeding. When he molts his train in summer, often in July, the breeding season is usually over, although some males will continue to mate successfully. One male can breed as many as eight females successfully, but five or six is more usual.

Flight pens need to accommodate the long train. They should be at least 6 feet tall and 12 feet wide, long enough to provide sufficient area for the number of birds confined. Plan on at least 80 square feet per bird, preferably 200 per male and 100 per hen. In cold climates, Green peafowl require more protection and heat than Indian peafowl. The India breed will manage cold weather without additional heat so long as it has protection from the weather, whereas the Green peafowl requires a heated shelter.

Roosts should be 4 or 5 feet above the ground and constructed with a flat surface, such as two-by-fours, rather than round. On flat roosts, the birds settle over their feet and keep their toes warm. In cold climates, peacocks' toes can get frostbitten if they roost on round roosts. Make sure to use untreated lumber. Toxic chemicals like arsenic and copper compounds used to make lumber rot-resistant and pest-resistant may leach into the birds through their feet over time and harm them. For the more delicate Green peafowl, roosts can be wrapped in electric heat tape and covered with carpeting.

Females begin laying eggs as early as February in warm climates, accumulating a clutch of seven to ten light-brown eggs laid every other day. Hens can lay up to thirty eggs if you remove them for artificial incubation or incubation by other sitters. They stop laying by September.

Eggs usually hatch in twenty-eight days but may hatch as early as twenty-seven days or take as long as thirty.

Peahens have camouflage plumage that protects them on the nest. However, peahens can display their feathers to defend their nests. Males and females have very different plumage, a common trait among ground-nesting birds. Male plumage is important in attracting a mate. Both sexes have a fanned crest. *Shutterstock*

Foster mothers, such as chickens, ducks, and turkeys, are good for brooding peachicks but are not as desirable as surrogate mothers. Peachicks are good flyers by the time they are a few weeks old, and they need to be enclosed.

Peachicks do well on game-bird or turkey starter with 28 to 30 percent protein and greens. If they are on range, chick starter can be adequate. Medicated starter can help them acquire immunity to coccidiosis. By the time they are eight to twelve weeks old, they should be eating lower-protein layer pellets, crumble, or turkey grower. The high-protein feed may cause slipped tendons if continued in their diet. Adults do well on game-bird mix. Dry dog or cat food is appreciated and provides protein. In cold climates, shelled corn generates body heat. Sunflower seeds are a nutritious treat.

Peafowl are omnivores and enjoy feeding on insects, small reptiles, and small mammals, making them effective pest-control birds. They enjoy fresh greens. They forage very well but may be difficult to attract back to the pen. Train them by letting only part of the group out each day and feeding late in the day to bring them home to roost.

In good health, some have been reported to live as long as thirty years, but twenty years is more usual.

PHEASANTS

There are thirty-five different species of pheasants, in at least eleven genera, with many color variations. As with peafowl, the relationships between the various genera and species are complex and not fully understood. Some, such as the Argus, are relatively unsociable, living solitary lives and encountering other individuals only to breed. Others, such as the Cheer and the Koklass, live in pairs. Still others, such as the Kaleege, live in colonies and breed in flocks.

Ring-Necked pheasants are such a successful import that many people believe they are native to the United States. They were first successfully established in the late nineteenth century in Oregon but soon ranged across the West and into the Midwest. The Conservation Reserve Program sets aside land that serves as a significant reserve of land-fowl habitat; this program could change as high corn prices create incentives to farm the land. *Shutterstock*

True or Typical pheasants are Common (Ring-Necked) and Green pheasants of the genus *Phasianus*. Other genera of pheasants include the Tragopans and Monals.

Vocalizations are important for all, bringing the solitary together and providing communication between pairs and trios and within large flocks. They also communicate with wing drumming, most highly developed in Kaleege and Silver pheasants. True pheasants are able to make a whirring sound with their wings. Wing drumming can be loud and intimidating to predators but also conveys communication between birds.

Pheasants prefer escaping from pursuers on foot, and most are not capable of much sustained flight. Reeve's pheasants are the best flyers, with reported flights of several miles.

Ring-Necked pheasants are the most popular pheasants hunted in the United States. Hunters flush them on foot or with dogs.

Pheasants are popular among birdwatchers. The Ring-Necked male's bright green head and dramatic markings are more frequently seen than the hens' drab camouflage coloring. Ring-Necked pheasants are about 20 to 28 inches long with a wingspan of 22 to 34 inches and weigh from 1 pound to as much as 6 pounds. Since they were introduced to the United States in the late nineteenth century, they have become established across the country.

The Ruffed pheasant, the Golden pheasant, and the closely related Lady Amherst's pheasant are among the best-known and easiest pheasants to keep and breed in a backyard aviary. In their native homeland of China,

This Ring-Necked pheasant hen shows the sexual dimorphism typical of ground-nesting birds, in which the sexes have very different plumage. Males defend harems of a dozen or fewer hens during the breeding season. Hens make nests in a shallow place on the ground. Provide plenty of vegetation for coverage in pens. *Shutterstock*

these pheasants live on mountainous slopes, ledges, and rocky hills often impenetrably covered with dense scrub, bamboo, bushes, and woods. At 36 to 40 inches, they are somewhat larger than Ring-Necks, although they weigh less. In the minds of many who breed pheasants, the Golden is unsurpassed in brilliance, beauty, and desirability as a pheasant for the game farm. They are hardy, disease-resistant birds. Both Golden and Lady Amherst's pheasants have been bred into color variations such as Tangerine, Silver, and Yellow Golden.

In the wild, pheasants may not be very wary. They are further disadvantaged by living on the ground in shrubbery, where predators may stalk them. Beebe has observed them partnering with flocks of thrushes in a reciprocal relationship, giving them a survival advantage. The thrushes feast on the insects stirred up by the pheasants, which then are alerted to danger by the calls of thrushes from their higher vantage points in the branches. Beebe also observed pheasants in reciprocal relationships with deer, which have an acute sense of smell. The pheasants have better vision, and the two groups mutually alert each other to danger.

The pheasant breeding season begins around March. Breeding groups should be penned together a month or so before. More than one male in a pen may result in fighting. A ratio of one rooster to seven or eight hens is good, although some have success with higher ratios. A hen lays about thirty-five eggs in a season, some as many as fifty. On their own, hens will incubate a clutch of six to twelve eggs. If you are going to incubate the eggs artificially, remove them at least once a day to avoid letting them warm in the sun. Store them in a cool place no longer than two weeks before initiating incubation. Pheasant eggs should be turned one or more times a day.

Lady Amherst's pheasant is closely related to the Golden pheasant. They often interbreed, and many specimens of both types are not pure. The name honors Sarah, Countess Amherst, wife of William Pitt Amherst, Governor General of Bengal, who was responsible for sending the first specimen of the bird to London in 1828. *Shutterstock*

The male Golden pheasant, with its colorful plumage, is a striking bird. Like other pheasants, the female is drab in color. Originally from China, Golden pheasants have been captive-bred since 1740. They are popular with fanciers and frequently seen at poultry and game bird shows. *Shutterstock*

Eggs hatch in as short as sixteen days for some species, twenty-three to twenty-six days for others.

You can also purchase hatching eggs or day-old chicks and start from there.

Pens must be covered with netting, or these birds will fly off. Ideally, a pheasant aviary has full-grown trees and shrubs for them to live in. If not, the birds appreciate a stack of cornstalks or other brush. Vulnerable birds are nervous birds and don't do well.

Commercial mixed-grain formulations of game-bird feed are satisfactory until you feel inspired to create your own mixture. Grit must be provided for proper digestion, unless the birds are housed on dirt that can provide it. Oyster shell provides calcium in the egg-laying season. Although not necessary, fresh greens are valuable. Birds with access to fresh green grass, kitchen and grocery store trimmings, or garden weeds breed better. They enjoy foraging for insects and grubs.

Keep feeders clean and make sure clean water is always available.

Pheasant poults can be raised intensively for food or game restocking. Chicks have plumage like their mothers until about two months of age, when males begin developing their colorful plumage. Birds reach maturity for table purposes at sixteen weeks of age. *Shutterstock*

QUAIL

The quail is the smallest of the gallinaceous game birds. All quail are small, plump birds that are not fully domesticated. They still have healthy populations in the wild, and tame birds revert easily to the feral environment. About sixteen species are found worldwide, loosely classified as Old World and New World. Old World quail, in the pheasant family, are the partridges of Europe and Asia. New World quail form a separate family that includes five species native to North America.

Quail are social birds, living in small family groups part of the year and joining with other families to live in coveys through the winter. Coveys may comprise hundreds of birds. As social birds, they have at least fourteen separate vocalizations to communicate with each other. You are likely to hear the distinctive "bobwhite" in the

Economic Impact

Birdwatching is more popular than ever, with forty-six million people watching birds—forty million of whom watch around their own homes and eighteen million of whom take trips to watch birds. Of course, they are watching more than gallinaceous birds, but the colorful game birds are always an exciting sighting. Waterfowl are the birds sought by 78 percent, and game birds like turkeys and pheasants make up 43 percent.

Birdwatchers spend $36 million on retail purchases, generating $85 billion in overall economic output and $13 billion in state and federal taxes, according to the U.S. Fish & Wildlife's 2001 survey. That translates into creating 863,406 jobs.

The survey showed that three million people age eighteen and over hunted game birds that year. The North American Gamebird Association reports that game-bird production and preserve industries have a $5 billion impact on the U.S. economy.

The National Wild Turkey Federation (NWTF) is active in developing habitat to support wild turkeys. Wild turkeys are a success story, now seen in fields and woodlands in every state. The population of turkeys, which had disappeared from many states entirely by the middle of the twentieth century, is now estimated at seven million.

The NWTF relocates turkeys to places that are appropriate habitat for them by supplying trapping equipment, transfer boxes, money, and volunteer assistance. The organization coordinates the relocation of some three thousand birds a year.

Bobwhite quail are benefiting from habitat restoration projects such as the Daniel Boone Ecosystem Restoration Project in the Daniel Boone National Forest in Kentucky. This native species has declined due to over-hunting and habitat loss. Private groups such as Quail Forever are partnering with government agencies to restore habitat and encourage bobwhite population increases. Removing invasive plants to allow native grasses to grow helps create habitats that allow quail to flourish. The native grasses grow tall in clumps and quail eat seeds and bugs under the overhanging grass. Introduced birds can supplement remnant native populations. *Shutterstock*

East and "cu-ca-cu" in the West as you hike through quail habitat.

Like pheasants, quail nest on the ground and roost above it, in trees and shrubs. Nests are not elaborate, but since they are made from surrounding grasses, they are very well camouflaged. Topped by a brownish hen, the nest is nearly invisible. Quail require early- and mid-stage successional plant communities, a healthy ecology of grasses and legumes, a wide variety of broadleaf plants, annual weeds, and brushy cover spread over the landscape. Fully grown mature trees and monocultures of grass and crops do not provide the low cover, seeds, and bugs that quail need for a proper diet.

Bobwhite, Blue Scaled, Valley (or California), Gambel's (or Desert), Montezuma's (or Harlequin, Mearn's,

Male California quail have more pronounced topknots, but females have their own. It looks like a single feather, but is six tightly packed feathers. California quail do not need to drink additional water during the wet season, as they can extract sufficient water from insects and moist vegetation they eat. *Philip Robertson, iStockphoto*

Painted, or Fool's), and Mountain Quail are native to North America.

Quail are territorial and may fight during the breeding season, so they are best kept as pairs or trios. Both parents set the eggs and raise the chicks. Eggs also hatch well in an incubator. A clutch contains twelve to twenty eggs.

Eggs hatch after about twenty-three days. Babies need to be kept in covered brooders from the start, since they are soon able to fly well enough that they may fly out of their container, where they can get chilled and die. Newly hatched chicks are up and out of the nest as soon as they are dry.

Button Quail

Chinese Painted or Button quail, only four inches long, are not truly quail but are instead related more closely to rails and cranes. They are often kept in botanical gardens and exhibits because they are effective at cleaning up seeds and bugs overlooked by larger birds. They acquired their name because they are "cute as a button."

Japanese Coturnix quail are the most common variety raised for meat. They mature early and may begin laying eggs as early as six weeks. Hens lay two hundred to three hundred eggs a year. Daylight duration of fourteen to eighteen hours a day is needed to maintain egg laying. *Shutterstock*

Pearl in the Egg

Back when my daughter, Nicole, was about ten years old, she found a quail egg on one of the paths between corrals at the boarding stable where we kept our horse. She was very excited about it and wanted to hatch it.

I figured the chance of it hatching was nil. Nicole found the egg right out in a public place, suggesting to me that a very young hen had been surprised when she laid it. It must not have been there long, and its prospects for hatching seemed wildly unlikely.

Nevertheless, I borrowed an incubator, and we sat and watched it. Quail eggs are supposed to hatch in twenty-three days. By the time we got to twenty-seven days, we figured we would have to give up on it.

A friend had come over to spend the night with Nicole, and the girls asked if they could take the egg out and open it, see what was inside. I agreed, since I was pretty sure nothing was in there.

A moment later, they raced back in the house, breathless. "It cheeped!" Nicole gasped. Sure enough, just as the girlfriend raised it up to smash on the walk, they heard the peeping. Nicole grabbed it and rushed it into the house.

We placed it in the incubator, and the next morning, we had a baby California quail. We named her Pearl, since she looked like a glowing pearl rolling across the bed. She snuggled up to us for warmth. As she grew, she often perched on the back of my neck, under my hair. Quail are social birds, and we were her covey.

She spent most days at liberty in the house. Quail droppings are small and dry, easy to vacuum up. She spent nights in the bird room outside in a cage. She liked to rest behind the television set. One night a possum threatened her, sneaking around, determined to get at her. She shrieked with fear, and we brought the cage inside.

She turned into a wonderful, dear pet whose company we enjoyed for about a year.

We were devastated the day she got out of the house while we were gone. We had actually gone out to see if we could find some other quail to live with her. She needed more social life than we could give her.

I feared a cat had gotten her. We put notices all around the neighborhood and in local vets' offices, but no sign.

Sitting down after lunch to read and nap, as was our custom, was not the same without her comforting presence. We cried, missing her.

A couple of weeks later, a woman called to say she'd seen our sign at the vet's office, and thought she knew where our dear Pearl was. The week before, she and her son had been astonished when a quail had emerged from the bushes in the preserve next door to their house and jumped right up on his arm, then jumped up and sat on his head! They often fed the wild quail there, but they had never had one do this. Quail generally keep to themselves. When she saw my notice, she put it together.

We were gratified to know that Pearl had found herself the ideal home: a regular food supply from humans but the large family she had always craved. I will always be grateful that she stayed with us so long and was such a blessing to us.

Chukar partridges are an introduced species that has adapted well to life in rocky terrain in North America's West. In its native range in Asia, birds living in humid climates develop darker and more intense plumage. Birds living in arid areas are grayer and paler. *Metzer Farms*

Rubberized non-skid shelf paper makes good footing for newly hatched chicks. It helps them to avoid the foot and leg problems that may occur on slick surfaces, plus it is washable.

Non-medicated game-bird starter spread on their floor helps chicks learn to peck for food. Greens, appropriately chopped up, can be introduced from the start, encouraging them to peck the food, not each other. As with other birds, overcrowding creates problems that can include pecking each other.

CHUKAR PARTRIDGE

This Eurasian partridge is a very adaptable dry-land bird. It has established populations in North America, where it has become a popular game bird. Its attractive markings are eye-catching.

In the wild, the Chukar can live on overgrazed land that will not support other birds. It breeds easily on its own, brooding eggs in a simple nest. Chicks are able to find their own food shortly after hatching, but both parents will spend time protecting them.

I couldn't resist getting a trio of Chukars once because they were so beautiful. They never tamed as much as other game birds did and persisted in escaping to the field across the road. Eventually, I gave up and let them live there. These sturdy birds have succeeded in occupying land that other upland game birds do not. They have taken an ecological niche ignored by native birds. They

are in demand for release and can reward the small-flock owner in the marketplace.

HEALTH MANAGEMENT

Coccidiosis is the most frequent infection for captive-raised game birds. Giving them medicated feed allows them to acquire immunity gradually.

Game birds are subject to all diseases that afflict other birds, including avian influenza. Always quarantine new birds for at least two weeks before introducing them to the flock. Peafowl are subject to intestinal parasites and should be wormed with a commercial product twice a year.

Keeping feed and water clean helps prevent infection. Feed and water dishes should be washed daily and dried thoroughly before refilling with fresh feed. Do not allow feed on the ground to become moldy.

Pheasants, quail, and partridge, including Chukars, can be incubated from eggs or purchased as chicks. Those intended for release to restock hunting reserves will need to be raised in flight pens after about six weeks of age. A flight pen 75 feet by 150 feet by 6 feet will hold four hundred to five hundred pheasants.

Peafowl need more space for their elegant trains. Flight pens should be much larger, as large as an acre. Breeding pens can be smaller, 12 feet by 18 feet by 8 feet.

Game birds do well on poultry feed or commercial game-bird rations.

Food plot mixes combining many of the crops that create habitat for game birds are available commercially or from organizations like Pheasants Forever. Seeds can be broadcast for easy establishment. Pheasants Forever produces Midwest Mix (corn, sorghums, sunflowers, and buckwheat), Nebraska Mix (a combination of sorghums and millets), and Western Mix (sorghums, sunflowers, millets, and clover). All are attractive to a wide range of wildlife. Select crops and maturities appropriate for your area, fertilize the plot, and control weeds to avoid excessive competition. Some weed cover benefits pheasants, but grain production will be reduced if weeds become a serious problem.

Weighing about a third of an ounce, quail eggs are popular gourmet products. Eggs can be boiled hard or soft, or pickled. They are also used raw in sushi. Shown here are quail eggs (*speckled*), chicken eggs (*brown*), and pigeon eggs (*white*). Pigeon eggs reflect the size of the breed and range in length from 1 to 2 inches and in diameter from ⅔ to 1¼ inches.
Arturo Limon, iStockphoto

PRODUCTS

Many of these birds are still hunted in the wild. Birds can be raised for release to restock hunting reserves. Peacocks are bred for preservation and exhibition as well as private aviaries. Conservation organizations emphasize habitat restoration over stocking domestically raised birds for hunting. Contact government wildlife agencies, hunting organizations, or Pheasants Forever for information on restocking.

Small flocks can be raised to produce meat and eggs for the table or local sale. Pheasants and quail varieties have been bred specifically for table use. Most of the quail raised for meat and eggs are the Japanese variety of *Coturnix*, an Old World quail. A Jumbo variety, weighing 1.5 pounds and dressing out at 8 to 9 ounces of meat, has been developed for meat production. Spanish and Italian varieties have been developed for eggs. Jumbo White pheasants, quail, chukars, and bobwhites offer more meat, and their white feathers produce the clean carcass appearance cooks want. Other table breeds include Alaskan Ring-Necked and California Buff Ring-Necked. White and Red Chukars have been similarly bred for table use. Jumbo Bobwhites and domestic varieties of Pharaoh Coturnix quail and Japanese Coturnix quail are available.

Peacock and pheasant tail feathers are in demand for clothing and costume decoration. Contact craft outlets and suppliers.

Marketing can be tailored to your local situation. Research what is in demand. You can succeed by filling a niche that is not being addressed. Check with local consumer outlets, such as retail food markets and farmers' markets. Talk to processors, grocery store managers, restaurant owners, and extension agents. They may have specific requirements for their needs. Meat inspection is not usually required, but check with the USDA Inspection Service and your state's Department of Agriculture for regulations.

White pheasants are often raised for table use. As with other fowl, the white feathers produce a more attractive carcass. Birds can be sold whole or as breast meat, frozen or smoked. *Shutterstock*

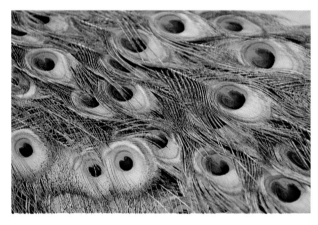

Peacock feathers are in demand for fashion and theatrical costumes. The iridescent colors shimmer and shine because of optical properties of the feather structure. The feathers have two reflecting surfaces over the pigments. Hues appear to change with different viewpoints because of the interference effects, which depend on the angle of light. *Shutterstock*

Pheasant feathers are sold for craft use, fly-tying, and clothing ornamentation, especially for hats. An Urdu expression translates to "She behaves as if she is decorated with a pheasant's feathers," meaning that she acts superior. *Shutterstock*

This Carrier pigeon is the type that made itself useful to humans in centuries past as a messenger. Homing pigeons are able to return to their homes from long distances, flying without rest for hours. Flights longer than one thousand miles are recorded. Pigeons probably rely on a variety of cues to guide them back to their home loft. Competitive contests compare Racing Homers on velocity, or the time taken to fly the prescribed distance. *Shutterstock*

PIGEONS

Pigeons are attractive, tractable, and easily managed by all ages in urban as well as rural locations. People keep pigeons for exhibition, sport, and meat production. The terms *pigeon* and *dove* are often used interchangeably. Both pigeons and doves are in the same scientific family, but there are hundreds of species. Divisions are not clearly delineated, but generally pigeons are larger than doves.

Keeping pigeons was probably one of the earliest forms of animal husbandry initiated by humans, as long as ten thousand years ago. Sumerian sculptures dating back five thousand years show doves associated with the goddess Asarte. The dove is associated with Noah, bringing back an olive branch to signal that the flood was over. In the New Testament, it is a symbol of the Holy Spirit, descending on Christ at his baptism (Mark 1:10).

Pigeon post is the use of homing pigeons to carry messages. Pigeons have the homing instinct, while doves do not. Pigeons have been used for that purpose since Roman times, when Julius Caesar used them in his con-

quest of Gaul. The ancient Greeks used pigeons to carry the results of Olympic contests. In the early nineteenth century, pigeon post was introduced into business and spread to military use by both the land and naval forces. Governments established their own lofts, and penalties for harming pigeons were severe. Pigeons were used extensively during both world wars. The British Dickin Medal, the highest animal award for valor, was awarded to thirty-two pigeons. The pigeon section was discontinued in 1950.

Pigeons are not native to North America. They arrived with colonists in the early seventeenth century. Escaped Rock pigeons became the feral birds that now inhabit cities so successfully.

Some doves are native to North America, such as the Mourning dove, but domesticated doves, such as the Ring-Necked, Laughing, and Barbary doves, are descended from wild North African doves. Doves and pigeons will breed naturally if confined together, but few of the offspring are fertile.

Pigeons and doves are also classified as either seed-eating or fruit-eating. Most birds kept domestically are seed-eating, but fruit-eating birds can also be kept successfully.

A pigeon loft can be any secure structure. Pigeons are usually allowed to fly free part of the time for exercise and training, although not all owners let their birds out, and not all pigeons even fly. Those that fly free need a way to get back into their loft but not get out again. Trap and bob entries allow pigeons in, then keep the door closed. Pigeons appreciate a flight cage, usually called a flypen, whether they get to fly free or not.

The loft should be separated into at least two separate coops: mated pairs settle better into nesting together if they are kept separate from the unmated birds. Unmated birds should be separated by sex, so that you can make breeding decisions. Young birds whose sex isn't yet obvious should be kept separate from breeding birds.

Pigeons and doves mate for life. Mated pairs may become territorial when nesting. You can further separate breeding boxes to avoid aggression between individual pairs. Even the pairs may need to be separated temporarily if the male becomes too demanding of the female. Aviaries with trees and brush are desirable for the hiding places and nesting sites that they offer, but nest boxes

Pigeon lofts can be plain or fancy. Pigeons are bred for behavior as well as plumage and conformation. Rollers perform backward somersaults in the air or on the ground; pigeon partisans consider them the Thoroughbreds of the air. *Shutterstock*

Pigeon hens lay two eggs in a clutch, but clutches overlap. Parents take care of two new babies in a separate nest as they manage the previous pair of squabs to adulthood.
Shutterstock

As meat birds, squabs are processed before fledging, when the meat is most tender. These youngsters have most of their adult plumage, but still show some baby down.
Shutterstock

can serve as well. The babies are called squabs. Two nests may be needed for breeding pairs, who may start a second clutch before the first squabs are fledged.

Pigeons are perching birds. Because they are territorial, they need a variety of places to perch. They can survive a wide range of temperatures and weather conditions but should have some protection from weather.

Pigeons lay two eggs at a time, and the parents share incubation duties. Eggs hatch in fifteen days for doves, eighteen days for pigeons. Hatchlings are atricial, meaning they are born naked and helpless, unlike ground-nesting birds. They are fed "crop milk," rich in protein and fats, which both parents make in their crops and regurgitate to their young. It is their sole food for the first few days. Squabs gain weight rapidly, doubling their weight in the first forty-eight hours and again in the next forty-eight. The parents gradually introduce partially digested grains mixed with the crop milk until the squabs are weaned at about three or four weeks old. The parents start laying in the second nest three to five weeks after the first squabs hatch, dividing their time between incubating the new eggs and feeding the squabs.

Pigeons will continue to breed successively throughout the year in warm climates, with time off for an autumn molt. They may raise as many as twenty-two young a year, although fifteen or sixteen is considered a good commercial production average. Fewer than twelve is not commercially viable. Squabs are taken from the nest for processing before they fledge, when their meat is most tender.

Pigeons begin to fly at four weeks of age. They are fully mature at six to seven months. Mature birds may breed well for five to six years but will continue to be productive for ten years or longer.

BREEDS AND VARIETIES

Like ducks and geese, pigeons are classified by weight into three classes: Light, about 8 to 10 ounces; Medium, about 16 to 17 ounces; and Heavy, up to 3 pounds. Wendell Mitchell Levi describes more than one hundred breeds in his *Encyclopedia of Pigeon Breeds*.

All fancy exhibition pigeons are varieties of the Rock pigeon. They vary in feather color, marking, and voice. Pouter and Cropper pigeons have enlarged crops that they inflate when cooing and strutting. All pigeons are able to inflate their crops, but these breeds have enhanced characteristics.

Pigeon feathers range across a wide spectrum of colors, patterns, and types. The blue bar shown here is the domestic version of the wild color pattern. Red, white, and black colors can be pied, grizzled, or splashed on the feathers. Fantails of domestic varieties have thirty to forty tail feathers, compared to the normal twelve to fourteen. Some have frilled or curled feathers. *Shutterstock*

Sporting or racing pigeons are kept for their flying or acrobatic abilities. They are shown in competitions. Rollers and Tumblers can do forward and backward somersaults in the air while flying.

Utility varieties are kept for meat but are also exhibited at shows. In some cases, two strains of a variety exist, show and utility, such as with King pigeons. Both are suitable for squab production, but the utility strain disregards the published *Standard* show requirements.

CHAPTER 9

RATITES

Ostriches, emus, and rheas (pronounced ree-ah) are all large, flightless birds of the group called ratites. Instead of the pointed breastbone of flying fowl, all the birds in this group have a flat, smooth sternum suggestive of a raft. The name *ratite* comes from the Latin word *ratis*, meaning "raft." Cassowaries and kiwis are also ratites, but they are kept in zoos and private collections, not as domestic farm species. Cassowaries in particular can have aggressive temperaments and require special handling.

Emus and rheas have three toes pointing forward. Ostriches have only two. This may be an adaptation for running, which they do very well. Ostriches and emus also swim well.

Ostriches, *Struthio camelus*, are presently native to Africa, although in the distant past they ranged across Eurasia and China. Like all wildlife, their population in the wild in Africa is much reduced due to hunting and loss of habitat.

Ostriches are curious and usually sociable. When raised around humans from the start, they are more likely to be friendly and less aggressive.

Ostriches do not adapt well to changes in their environment. It's best to start them where they are going to remain. If they must be moved geographically, one trick is to keep their bedding the same. Otherwise, they may eat indigestible material, and it can become impacted in their stomachs. Ostriches peck between two thousand and four thousand times a day.

HISTORY AND CULTURE

Ostriches are mentioned in several places in the Bible, with modern translations changing *owl* to *ostrich* in references in Job (30:29), Isaiah (13:21, 43:20), Micah (1:8),

Domestic ostriches have been ranched for more than one hundred years, although they were not included in the USDA's Agricultural Census until 2002. The Farm Bureau recognizes ratites as a commodity and includes industry representatives in its commodity meetings and reporting. *Shutterstock*

134

and Jeremiah (50:39). The correction is based on the contexts being consistent with ostriches, not owls. Job discusses the ostrich with God (39:13-18), including references to indifferent mothering and running faster than the horse. Ostriches can run as fast as 40 miles per hour in short bursts and can maintain a steady speed of 30 miles per hour, making them the fastest two-legged animal in the world. Ostriches are mentioned in Leviticus (11:13) as unclean, along with eagles and ospreys. Ostrich meat is not kosher. Lamentations (4:3) presents the ostrich as an example of cruelty for not being a good mother. Although that accusation has been made against them, Brian Bertram, as reported in *The Ostrich Communal Nesting System*, found the situation more complex when he studied ostriches for three seasons in Tsavo West National Park in Kenya. He observed that although several females lay their eggs in a single nest, only one female will partner with the male in incubating the eggs. That female is able to recognize her own eggs and may nudge the eggs of other birds out of the nest if it becomes crowded.

Ostrich feathers have been associated with nobility and authority since the days of the Egyptian pharaohs. Queen Elizabeth I in the seventeenth century and Marie Antoinette in the eighteenth century adorned themselves with the aristocratic feathers. Their feathers came from wild birds in North Africa. Feathers were not imported from South Africa until the nineteenth century.

The goal of improving feathers for the feather market drove the development of the hybrid domestic ostrich, the South African Black. Between 1886 and 1910, wild Barbary birds from North Africa, with their fuller and closer feathers, were bred with tame Cape birds. However, the market for those improved feathers collapsed in 1914. World War I, changes in fashion away from large-plumed hats, and an anti-feather campaign to protect wild birds from slaughter combined with overproduction and disorganized marketing to produce disaster. Ostrich farmers abandoned the business and, in some cases, their birds. In recovering from that setback, a group of South African farmers organized the Klein Karoo Agriculture Co-operative in 1945. The group advanced the interests of their members while attempting to control all ostrich marketing worldwide.

Ultimately, eggs were smuggled out of South Africa to Israel in the 1970s. The resulting birds became the African Black breed, as they are known throughout the world except in South Africa today.

Emus are the national bird of Australia, where wild ones are legally protected. All farmed emus are domestically hatched and raised. Two species of small emu became extinct shortly after European contact in the eighteenth century, but the large *Dromaius novaehollandiae* remains common in the wild. Emus on Tasmania also became extinct after European settlement.

Aborigines hunted emus for meat, oil, and feathers. They hunted them actively with spears at waterholes, where they would set out a preparation made from the pituri plant, which drugged the birds and made them easy to spear. They also dug pits rigged with spears to impale birds that fell into them, and they snagged emus in nets. The oil was rubbed on their skin for medicinal purposes and mixed with ochre pigments to make the traditional paint. The oil was also used to lubricate wooden implements. The feathers were used in many practical and decorative ways, including as a ring pad, or *akartne*, to cushion heavy items carried on the head.

Natives in southeastern Australia credit the emu's egg with being the origin of the sun. According to mythology, after Dinewan (the emu) and Bralgah (her companion) argued, Bralgah took one of Dinewan's eggs and threw it into the sky. It broke on a pile of firewood and burst into flames. The deity in the sky saw that the light helped the

The Cassowary is a ratite but is not commercially raised. They do breed in captivity and are kept as display birds in private collections. *Shutterstock*

Ostriches occasionally raise their wings and spin around. This behavior is apparently an expression of exuberance. They may get so dizzy they fall down.

Theodore Roosevelt witnessed this behavior twice in the wild. It was his opinion that the ostrich didn't risk drawing extra attention from its predators by such a conspicuous display. "By the time the young birds are old enough to gyrate or waltz, they are so conspicuous that any foe is sure to see them, whether they are walking about or gyrating; and after their early youth ostriches do not seek to escape observation—they live under such conditions that they trust exclusively to seeing their foes themselves, and not to eluding the sight of their foes," he wrote in the *Atlantic Monthly* of June 1918.

world and gathers firewood every night to continue the fire burning during the day.

Rheas are South American members of the ratite family. The American rhea, also called the Gray or Common rhea, is the largest American bird, at a height of 4 to 6 feet and a weight of about 50 to 80 pounds. A smaller species, the lesser or Darwin's rhea, is not raised commercially.

Farmers have been known to shoot rheas if they find them eating their crops. The birds are partial to broad-leafed plants like cabbage, chard, and bok choy and avoid grasses unless there isn't anything else to eat. Egg gathering and loss of habitat, especially due to crop burning, have reduced their population in the wild. The American rhea is a threatened species listed in CITES, although it is only near-threatened according to the IUCN.

OSTRICHES

Five subspecies are identified in the wild, but all commercial domestic ostriches are the Domestic or African Black. Both sexes have a bluish-gray neck. This breed is shorter, smaller, and darker than its wild cousins.

One domestic ostrich egg weighs 3.5 to 5 pounds, equivalent to two dozen chicken eggs. Hens lay an average of fifty eggs per year, with a range between thirty

Ostrich eggs, shown here next to chicken eggs, vary in size from as small as 16 ounces to over 4 pounds, averaging about 3¼ pounds. They are different in shape, less pointed than chicken eggs, and may be almost spherical. The shell is two millimeters thick.
Gillet Luc, iStockphoto

to eighty eggs. They are seasonal layers, with laying beginning with longer days in the spring. If allowed to incubate the eggs naturally, the hens will lay in a collective nest, which they will set on during the day and the male will set on at night. The incubation period is forty-two days.

The largest birds in the world, adult ostriches weigh 200 to 350 pounds and stand 7 to 8 feet tall. Market birds are typically slaughtered at twelve to sixteen months of age, weighing at least 225 pounds.

Ostriches are territorial and may be unexpectedly aggressive. They signal territorial behavior by hissing, stamping feet, flaring wings, charging, and kicking. They can inflict serious injury and even kill a person with those large legs. If confronted with an attack, lie down and wait for the bird to abandon the area. Not for the faint of heart!

Ostriches use their wings in displays that communicate to other ostriches. Males perform courting, aggressive, and threat displays. Females demonstrate submissive behaviors, and males use their wings to invite the females to the nest site. Both males and females spread their wings and wave their necks to distract predators, including humans, from their chicks. The wings also serve to lower body temperature, counterbalance while running, protect the eggs, and shoo away flies. *Shutterstock*

Docility and manageability are important considerations in acquiring birds. It's important to learn techniques to manage them for treatments, leg band adjustments, and other intimate procedures. Ostriches can be allowed to stand for many procedures. Smaller birds may have to be turned over. Hooding them can make them calm and easy to manage.

EMUS

Emus are slightly smaller than ostriches, reaching 5 to 6 feet tall and 90 to 140 pounds. They adapt well to both hot and cold temperature extremes. They enjoy playing in water and mud and can swim. Emus are more docile than ostriches and rheas. They are adapted to travel long distances at a walk and trot and can run in bursts as fast as 30 miles per hour.

Three subspecies, separated geographically, are identified. The only wings they retain are vestigial. Their leg muscles make up for the lack of flight muscles. They make loud booming, drumming, and grunting sounds that carry farther than a mile. Their shaggy brown feathers insulate them from the heat, and they also pant to keep cool.

Emu feathers are different from other bird feathers. The barbs on the feathers are widely spaced so that they

Dark green emu eggs range in size from 1¼ pounds to 1½ pounds, ten to twelve times the size of chicken eggs. They cannot be visually candled. Infrared candling tools have been developed to determine whether an embryo is developing by locating the size and shape of the air cell. Experienced breeders tap the egg with a metal rod, judging whether an embryo was developing by the sound produced. *iStockphoto*

do not interlock. Lacking that connection, they resemble hair more than feathers.

Males build the nest and incubate the eggs, which take fifty-two to fifty-six days to hatch. The dark green eggs are difficult to candle, but infrared candlers are available.

Emu are well adapted to their hot, dry native Australian environment. Their feathers look shaggy, but they insulate the bird against heat. The loose feathers protect the skin and body from absorbing too much heat. In hot weather, they pant to maintain a comfortable body temperature. *Shutterstock*

The emu's dark green eggs take more than fifty days to hatch. Joanne Rigutto of Oregon sets them every few days in her incubator, to stagger the hatch. She sees the egg move before pipping starts and finds that, at that stage, they will chirp and whistle in response to her whistle. Pipping begins within a day. *Joanne Rigutto*

Joanne observes the emus closely once they begin to hatch and helps them out if they don't succeed in freeing themselves within a day. Spending longer than that in the shell can cause leg and spine problems. Helping them out before they are ready, however, can result in the chick bleeding out through the umbilical cord. *Joanne Rigutto*

The chicks rest in the bottom incubator tray for a day. This gives them time to dry thoroughly and the umbilicus to close. Because emus hatch in the winter, Joanne sometimes keeps them indoors until the weather warms and the chicks have enough feathers to maintain their body temperature and withstand conditions. She hardens them by raising the heat lamp over the tub a few inches every few days, until they no longer need the lamp day or night. *Joanne Rigutto*

Joanne lines the bathtub with carpet to give the birds' feet some security and to avoid developing leg problems such as splay leg. Placing feed at one end and water at the other encourages the birds to walk back and forth for exercise. When they are ready to go outside, she watches them and brings them back indoors if they start to shiver. They are curious and enjoy exploring new situations. "One of a young emu's favorite things to do is to run and dance!" she says. *Joanne Rigutto*

Females lay as many as twenty eggs in the season. They are winter breeders, stimulated to breed by the shorter days of October through April in the Northern Hemisphere. In their native land, they breed from April through September. Eggs weigh 1 ½ to 2 pounds and are equivalent to about a dozen chicken eggs.

Chicks begin life by eating the egg from which they hatched. They have striped plumage, which makes an effective camouflage. They lose the stripes by the time they are two years old.

RHEAS

Rheas stand five feet tall, the tallest American birds. Commercially, 95 percent of the bird is considered usable as meat, feathers, oil, and leather. They are less domesticated than ostriches and can be aggressive to humans and other birds.

Rhea eggs are yellow, but the color can bleach almost to white in the sun. The rhea laying season north of the equator is from May through August, during which time females may lay from twenty to sixty eggs. The female

The rhea nest site is selected by the male. He usually digs a shallow depression with his feet and lines it with any available material. He attracts the females with a booming vocalization and wing display. His mating call is like a fog horn. Some males vocalize year-round. Males may be aggressive toward humans, especially during the mating season. *Craig Hopkins*

The male rhea cares for the eggs, turning them as needed. He may be aggressive even to the female when she comes to the nest to lay another egg. She sits beside him to lay her egg, which he then rolls into the nest. Hens usually lay every other day. *Craig Hopkins*

The male rhea decides how many eggs he will incubate. If the site is subject to disturbance, he may attempt to move the eggs or abandon the site. Providing a blind for your Rheas encourages them to feel secure. *Craig Hopkins*

Some male rheas remain aggressive year-round, but others resume more docile behavior after the breeding season is over. Wings help cover the eggs, although this one required some adjustment after the photographer was chased away. *Craig Hopkins*

White and gray rhea chicks hatch out of their eggs in a renovated redwood incubator. Craig Hopkins of Hopkins' Alternative Livestock in Indiana replaces electrical and mechanical parts to give new life to old equipment. The birds will be introduced to grass and clover pasture within the first three days of life. They benefit from exercising their legs right from the beginning, as they would in the wild. "We have seen one-day-old chicks walk one to two miles while being raised by their father," Craig says. *Craig Hopkins*

Rhea chicks get started on chopped leaf lettuce and ratite starter. Grit is always available. Craig Hopkins uses surplus males to help raise chicks on his farm. He incubates eggs artificially until a few days before hatching. They are then placed in a nest previously prepared by allowing the male to set on dummy eggs. The chicks and the adoptive father bond by calling to each other prior to hatch. Additional incubator-hatched chicks can be added later. Male rheas will adopt chicks, even attempting to adopt chicks from other families if they are pastured together. *Craig Hopkins*

lays her eggs by squatting close to the male. After she lays the egg, the male rolls it into the nest he has scraped in the ground. The eggs go into a collective nest, which is then incubated for thirty-two to forty-two days. The male is responsible for all incubation and chick care. He leaves the nest only to eat, drink, and breed.

FEEDING

Ratites have not attracted as much scientific attention to their nutritional needs as commercial poultry have. Some universities and cooperative extension services have taken an interest in ratite nutrition. The birds in this category differ substantially from each other, making generalizations difficult.

The ostrich and rhea have the digestive systems of herbivores, but the emu's system is more like that of an omnivore. Ostriches and rheas have long ceca, which are sections of the digestive system used to ferment grass, to allow them to extract nutrition from fibrous grasses. Consequently, their feed takes much longer to pass through their bodies—thirty-six to thirty-nine hours—than the five to six hours it takes food to pass through emus. Ratites do not have a crop as other domestic birds do. In ratites, the gizzard, also called the second stomach and the ventriculus, contains the grit and performs the role of grinding the feed and mixing it with gastric juices.

The proventriculus or glandular stomach is a true stomach, the initial organ of digestion. It may become

Grit

Ratites, like all other birds that lack teeth, need appropriately sized grit to digest their food. Granite chips, gravel, or smooth stones will work. Optimal size is half as big as the bird's toenail. Milled feeds made of alfalfa, soy, and corn supplemented with vitamins are softer and easier to digest than the forage birds eat in their natural habitat. Ranch-fed birds may not need grit to get nourishment from their food.

"If you get the right feed, then you don't need grit," said Steve Warrington, CEO of Ostriches On Line.

impacted if too much indigestible material or dry grass is consumed. Clean water must always be available to moisten the food. Ratites will refuse to eat if water is not available. Ostriches drink about three times as much as they eat.

Experienced owners, extrapolating from other poultry and the behavior of these birds in the wild, have arrived at successful feeding and exercise regimens. Major feed companies offer general ratite formulations. Different commercial formulations may be available for each type. Work with the feed store to locate them and order a regular supply.

Duck or game-bird starter/grower feed, with about 20 percent protein, can be used for chicks if specialty feeds are not available. Protein content should be reduced after two or three months of age until they are getting 13 to 15 percent protein by six to ten months of age. Ostriches grow fast and can develop leg problems if they get too much protein. Nutritional imbalances can make rapid-growth problems worse.

Ratites are often raised on pasture. They particularly enjoy semiarid pasture that resembles the vegetation of their native habitat. However, most pasture does not offer sufficient nutrition, and they will need supplemental grain. Ratites are rapacious eaters and will devastate pasture, so be sure to rotate their enclosure.

Chicks need to establish healthy digestive bacterial flora. The administration of probiotics is preferable to feeding chicks feces from other birds, which can give the young birds diseases for which they have not yet acquired immunity.

HOUSING

Ratites need strong fencing, at least 6 feet high, and preferably 8 feet. They can easily jump higher than 5 feet. Fencing should have a smooth, solid top rail and be easily seen, so that they don't become entangled in it. Fencing should have no sharp edges that can cut or tear skin. The hide will scar and lose value as leather.

They must have adequate room to move around. These are birds that, in the wild, travel many miles every day. Lack of exercise may be the cause of leg problems in growing birds.

Ostrich chicks start out needing about 13 square feet of shelter and 16 square feet of pasture per bird, increasing as they grow to 50 square feet of shelter and more than 2,500 square feet of pasture for pre-breeding adults. Breeding birds require more than 10,000 square feet per bird.

HEALTH MANAGEMENT

Adult ratites are very hardy. Thus far, no diseases have been documented to affect emus. They live as long as thirty years.

Most mortality occurs in the first three months of life, and most of those losses are due to management

Shelter

Ratites adapt to a wide range of temperatures and climatic conditions, but they should have at least a three-sided shelter to protect them from storms. They need shade from the sun. Existing outbuildings may be remodeled to accommodate them. Metal-frame buildings can work well.

Set food and water in a holding pen area separate from the rest of the pasture to help get the birds' cooperation when you need to capture them. They will be accustomed to congregating there.

Most chicks are hatched in incubators and must be kept warm in a separate brooder facility until they are able to regulate their own temperature. They can be allowed outdoors in warm weather when they are three days old.

Like other birds, ostriches enjoy a good dust bath. They toss the sand onto their feathers, rub it around, and finish up with a good shake. Working dust into the feathers removes stale, oxydized oil, which birds then replace with fresh oil by preening. A bird preens by dipping its beak into the oil produced by the uropygial gland at the base of the tail and spreading it on the feathers. Birds show obvious pleasure in dust bathing and may spend fifteen minutes or more at it. *Shutterstock*

factors. Chicks may be born with congenital deformities or develop problems with the yolk sac. The yolk sac should be completely retracted into the body when the chick hatches. Allow the navel to close before removing a chick from the hatcher. The chick will continue to receive nourishment from the yolk for as long as five days, but it should be encouraged to eat immediately. External yolk sacs require surgical intervention from a medical professional.

Chicks may develop scoliosis or deformities of the long leg bones. Leg problems can develop literally overnight. Splinting the leg, if done correctly as soon as the problem appears, can resolve it successfully. Get an expert.

Impactions of feeding material in the proventriculus are a risk whenever feed or bedding is changed. These birds are curious and will ingest almost anything. Avoid lining the pen with anything that unravels or comes apart, such as carpeting. Chicks are particularly at risk for acute impactions from eating indigestible material and can die without prompt surgical treatment. Chronic impaction develops from a partially obstructed proventriculus that allows some food to pass. It may become infected. William Sutton, M.D., of Chulagua Ranch in Texas, writes, "They love a stable environment. They love boredom."

No medications are licensed for use in ratites, so any prescriptions are made "off-label," according to the

Allen and Myra Jones Charleston compiled this list of plants toxic to emus from their experience with the birds. Myra is an At Large Director of the American Emu Association.

This is certainly not a complete list of plants dangerous to emus, but it is a start. Please keep in mind that these are plants found in the continental United States.

- **Any of the Buckeyes** (California, Ohio, red, painted): The seeds cause muscle weakness, paralysis, dilated pupils, diarrhea, and stupor before death.

- **Cherry** (sweet cherry, wild cherry): Seeds, leaves, and twigs cause gasping, respiratory failure, spasms, convulsions, and death.

- **Black locust:** Young leaves and seeds cause weakness, dilated pupils, bloody diarrhea, and weak pulse. The circulation slows and legs get cold; birds go into shock. Can be toxic if large quantities are eaten.

- **Buffalo nut:** Seeds cause severe irritation of the mouth similar to that caused by blister beetles.

- **Chinaberry:** Fruit in large quantities causes stomach irritation, bloody diarrhea, paralysis, irregular breathing, and respiratory distress. There are incidences of death in chicks.

- **Hydrangea** (mountain, French, peegee, oakleaf): Can be toxic to chicks but will be toxic to adults only in large quantities.

- **Jimsonweed** (Jamestown weed, thorn-apple, stinkweed, datura): All parts of this vine are toxic. In humans it causes hallucinations; emus reportedly go wild, so we can only guess. Also causes pupil dilation, erratic pulse, sometimes stupor, and convulsions before death. Large quantities are required to kill an adult emu.

- **Kentucky coffee tree:** It is reported that chicks eating seed pulp may die. Adults are not affected as severely, but it may be toxic if enough is eaten.

- **Lambkill** (sheep laurel, wickey, swamp laurel): All parts are poisonous and cause lack of coordination, paralysis, convulsions, and death.

- **Laurel** (Carolina cherry, English): Leaves, twigs, and stems cause gasping, weakness, excitement, pupil dilation, spasms, convulsions, respiratory failure, and death.

- **Oak:** Acorns cause bloody diarrhea and excessive thirst and urination.

- **Peach:** Pits from fruit, stems, and leaves are reported to cause convulsions, respiratory failure, and death.

- **Pigweed** (rough, green, winged): All parts are poisonous and cause lack of coordination, paralysis, convulsions, and death.

- **Soapberry:** This reportedly kills chicks if the fruit is eaten in large quantity.

- **Flowering tobacco** (nicotiana): All parts are poisonous and cause diarrhea, staggering, breathing difficulties, and death.

- **Yew:** Leaves and seed pits are reported to cause trembling, difficulty breathing, diarrhea, sometimes convulsions, and death if large quantities are eaten.

veterinarian's knowledge and experience. Medications can be given orally, and many birds will peck at brightly colored pills willingly. Birds being raised for meat must be removed from medication at least thirty days before slaughter.

PRODUCTS

Ostriches, emus, and rheas all are being raised in the United States for their meat, leather, feathers, and other products.

Ratites have had a boom-and-bust economic cycle. High prices for feathers fueled an initial boom in the late nineteenth century, when ostriches were first introduced in the United States. The market declined in the twentieth century with international conflict and changing fashions. It didn't recover until the late twentieth century, as ostrich meat and leather gained visibility and commercial emu farming got its start. Prices for breeding birds soared, then collapsed when anticipated market demand did not develop. Many producers took their losses and withdrew from ratite farming.

The products are desirable, and if a reliable market can be established, ratites have the potential to provide income to small producers. Price is an obstacle for the consumer, with ostrich steak selling at around $20 a pound. Research this thoroughly before investing money.

The meat is considered comparable to beef in taste but lower in fat, lower even than chicken. Meat birds must be processed under USDA inspection. Steaks, fillets, medallions, roasts, and ground meat are available for the table. Farm-raised meat is usually tender and does not require slow cooking. Because of its low fat content, it can easily be dried out and overcooked.

Ostrich feathers have been in demand for thousands of years. In the twenty-first century, they are used mainly for ornamentation. They can be harvested from live birds by plucking and clipping as well as at slaughter. Emu feathers and toenails are also used in decorative arts and crafts.

Ratite leather has a distinctive pattern from the quills. Eggs are used in art projects. Rhea eggs are smoother and easier to carve than the dark green emu eggs and larger, harder ostrich eggs.

Ostrich leather is a high-end product used to make boots, handbags, wallets, belts, and clothing. Many designers select it for its unusual quill pattern. It is very durable, and its natural oils help it resist drying and cracking. The "Certified American Ostrich" trademark identifies a Quality Assured Ostrich product based on ostriches that are hatched, raised, and processed in the United States under specifications set up by the American Ostrich Association (AOA). All leather labeled "Genuine American Ostrich Leather" must be from birds that are source-verified, and hides should meet the standards set by the AOA Hide Guide and appear as healthy without signs of distress. *Shutterstock*

Emu and rhea oils, rendered from the fat, are marketed as healing and cosmetic products. Some clinical evidence indicates that it is anti-inflammatory and acts to desensitize the skin, but the Food and Drug Administration has not confirmed those claims.

Marketing organizations, such as the American Ostrich Association, the American Emu Association, and the North American Rhea Association, exist to support ratite farming. Ostrich.com (U.S.) and Ostriches On Line (U.K.) provide technical and business information to develop an ostrich farm, including special software to create a business plan.

Check with your state department of agriculture, since some states classify ratites as exotic animals that require special permits.

MANAGEMENT

Guinea fowl are companionable with other poultry in the barnyard. They may even lay eggs in each other's nests and incubate each other's eggs. It's a Go Along and Get Along world. *Shutterstock*

The various poultry species differ in many ways, but they also share many commonalities. General husbandry and management principles and practices can address their common needs. Always be aware that they do differ, and check out what applies to your individual situation.

MIXED FLOCKS

Keeping birds of different species together often works out, provided you are aware of the pitfalls. Consider your own circumstances of space and facilities.

The various species differ in their vulnerability to diseases. For instance, histomoniasis, a protozoan infection, becomes a serious problem for turkeys, but it's less of a problem for other species.

Ducklings and chicks can be kept together in terms of temperament, but ducklings will be safe in water that could drown chickens. Generally, it's better to separate them. Ducks and geese are usually safe together. Sufficient area to avoid crowding is always important. Crowded birds will fight. Do not house

Waterfowl, both domestic and wild, will share a pond. Mute swans are considered undesirable invasives in some areas. Check local and state regulations for guidance. Reports of migratory waterfowl carrying Highly Pathogenic Avian Influenza have not been confirmed despite an intensive government monitoring program. *Shutterstock*

Crucial factors for successful hatching in artificial incubators are:

- Temperature
- Humidity
- Ventilation
- Egg turning
- Duration

Each poultry species has an optimum temperature for hatching and a different length of incubation, so research the kind of eggs you plan to hatch. Temperatures too low will not support development of the embryo, and too high will oppress it.

Too little humidity dries out the embryo, leaving it weak and sickly. Too much makes the embryo too chubby and sticky to escape from its shell. Eggs of different sizes lose moisture at different rates, creating problems for incubation of eggs of different species in a single batch.

The embryo "breathes" inside the egg, and exchange of carbon dioxide for oxygen is essential. Fresh air must be circulated in the incubator without drying or cooling the eggs.

Eggs are turned to move the embryos around and prevent them from sticking to one side of the egg. Most species require turning more than once

Shutterstock

a day until the last few days before hatching. Large incubators do this automatically. In smaller ones, the eggs must be turned by hand.

Commercially built incubators are available to hatch a few eggs or many. You can make a small one for yourself out of a Styrofoam ice chest, or build a larger one. Small ones depend on con-vection to move stale air out and draw fresh air in. Larger incubators have fans to ensure air exchange. Follow directions for best hatching.

peafowl with other birds. They are vulnerable to diseases that chickens can carry.

Male guinea fowl will defend food and water against chicken roosters, so do not keep them confined together in a common yard. They can be cooped together at night, but the roosters will not survive if their lives depend on sharing the yard with guinea fowl males. Guinea fowl hens get along fine with chickens and roosters, and occasional hybrids result. They are called guin-hens.

Chicks, ducklings, goslings, cygnets, and other baby birds form a strong social bond with their parents, or whoever fills that role, by imprinting early in life. The attachment helps them to form cohesive families that are more likely to raise young successfully. However, if surrogates take the place of same-species parents, young birds will identify with them, birds or animals not of their own species. When that happens, they may never be able to mate or form bonds with members of their own species.

Check for embryo development and for cracks by candling eggs. This one shows an egg with no embryo. Within a few days of incubation, the dark form of an embryo will be visible. If not, discard it to avoid rotten eggs exploding in the incubator. A flashlight with a cardboard tube attached is adequate for small numbers. Don't set or sell cracked eggs, as cracks can allow bacteria to enter; instead, boil them and feed them to your birds or other livestock and pets.
Metzer Farms

BIOSECURITY

Communicable diseases can be stopped at the front gate by adopting biosecurity practices. Having a secure program in place may also help you advocate for the survival of your flock against mandatory culling in the event of a disease outbreak in your area. Keep records of the measures you take and any new birds introduced to your flock. Diseases can be spread by dirt, feathers, and any other items that carry infection with them, as well as by live birds. Avian influenza is the most notorious, but birds can share many other diseases.

Acquire birds only from sources that can verify they are disease-free. Then quarantine new birds for two weeks in separate quarters to watch for symptoms and ensure that they are healthy.

Keep boots dedicated to working inside the pen or pasture. Leave boots or shoes that may be contaminated away from contact with the birds. Jackets or overalls used only for contact with the birds may also be used to prevent contamination. Boot covers can be used. Disinfectant foot baths can be used at entry and exit. Visitors must observe disinfectant procedures.

Wash your hands before and after handling birds or anything that goes into contact with them.

Waterfowl Cautions

- Raise young birds on rough surfaces, such as corrugated plastic shelf liner, cloth, or burlap, that give them traction. Slippery surfaces like newspaper can cause leg problems.

- When using wire mesh, make sure it is small enough to support tiny legs. You can use ⅜-inch mesh with larger breeds.

- Keep litter dry. Remove litter before it compacts or gets moldy.

- Keep young domestic waterfowl away from swimming until they are a month old. If they get chilled, they are subject to illness or may crowd together for warmth and accidentally crush each other.

- Litter chips must be large enough that young birds aren't able to eat them. They can die from eating litter. Absorbent material like chopped peanut hulls, pine shavings, and straw work well.

STAY HEALTHY

Keeping your flock in good health is the best defense against disease. Keep them clean and well nourished, and you will avoid most diseases. Any diseases they do catch will be less severe and their recovery faster.

Clean, fresh water and food reduce the opportunities for spreading disease. Dirty water breeds pathogens. Many birds will avoid dirty drinking water, which in turn subjects them to stress of dehydration. They will also stop eating if they do not have water to drink, making them more vulnerable to illness.

Avoid crowding birds. Give them plenty of room to roam. Crowding causes stress, which reduces their ability to resist disease or recover if they do catch something. Crowding also increases the speed and reach of disease spread. Birds that have more space are less likely to catch diseases from each other.

Control rodents in the housing and yards. They can carry diseases and parasites.

Fresh air and sunshine are natural disinfectants. Keep the poultry yard well drained. Even waterfowl need dry places to groom and sleep. Most poultry enjoy dust bathing.

When using any chemical treatment—whether it be an antibiotic or other medicine or an insecticide—read and follow all package directions.

AVIAN DISEASES

Avian influenza has many forms, and all birds are subject to them in varying degrees. The highly pathogenic (HP) forms like H5N1 are rare. Occasional infections of H5N1 in humans have raised the specter of a mutated virus that could be transmitted between humans. Human cases remain rare, a few hundred worldwide since human infection was first identified in 1997. In comparison, common forms of influenza kill around 500,000 people annually worldwide. Nearly all of the 200 million birds that have died have been killed by official culling rather than the disease itself, according to Elizabeth Krushinskie, DVM, PhD, vice president of Food Safety and Production Programs at the U.S. Poultry and Egg Association. H5N1 has not yet reached the American continents and may never arrive here.

The H5N1 strain of HPAI kills most chickens that catch it, but ducks may be infected without dying or even getting very sick. Wild geese and swans also catch it, and because wild populations migrate, this led to fears that they could spread it, but predictions that migratory waterfowl would carry it across borders have not been supported by evidence. The U.S. Geological Survey has not substantiated claims that migratory waterfowl spread highly pathogenic avian influenza, despite sampling more than 190,000 birds as of October 2007 to identify it. Ornithological organizations like Birdlife

A mixed group of turkeys gets along well, but they will have to be separated for breeding, otherwise varieties crossbreed and you won't be able to tell what you have. Breeding traditional varieties can be a significant marketing advantage. Breeding for exhibition also requires selective breeding techniques. Deliberate crossing can result in new color varieties, but segregating breeding pairs is necessary to be certain which varieties have crossed. *Shutterstock*

Small wetlands are good for waterfowl and bring other benefits to your land. These environments help clean the groundwater and reduce flooding while improving habitat for other wildlife. Even a small property of one or two acres can accommodate a pond.

Thomas R. Biebighauser, wildlife biologist for the U.S. Forest Service Center for Wetlands and Stream Restoration in Morehead, Kentucky, advises property owners on building small wetlands. A deep pond isn't necessary. Small, shallow areas can make wetlands that ducks, geese, and swans can easily access.

Wetlands do not need to be expensive. Get soil tests first, to determine what your site will require. Soils that are high in clay will retain water naturally. Quality synthetic, aquatic, fish-grade liners can be used in more easily drained sandy soils. A pond 30 feet in diameter with a gradual slope to a depth of 18 inches can be bulldozed for around $1,000. Erosion control can be managed by planting fast-sprouting grains like wheat and rye. Bed with straw or mulch, not hay. Hay introduces too many unwanted seeds, and you'll have a weed problem later.

U.S. Fish and Wildlife Service

Tom Biebighauser

Tom Biebighauser

Sedges, rushes, and wildflowers will grow in naturally. "Native plants will come in with every duck and bird," says Biebighauser. "Within five years, you will have fifty species." Adding native plants and seeds speeds up the process and increases diversity.

Soon, dragonflies and frogs will move in. Mosquitoes won't be a problem so long as the critters that feed on them are allowed to flourish. Dragonfly and salamander larvae eat mosquito larvae and eggs. Encourage them by getting water from a clean source like rain or natural springs. Don't introduce fish, which will eat the beneficial invertebrates and compete with your birds for food. "If you have fish, you won't have those water striders," says Biebighauser. "Mosquitoes check in but won't check out."

Design the wetland so that it will dry up once a year. Drying out cleans it.

Wetlands will attract all kinds of wildlife, including those that prey on domestic birds. Provide a floating platform for nesting and retreat. Lock birds up at night.

"It can be doable," Biebighauser says. "Make your farm wildlife friendly."

International and the British Ornithologists Union have analyzed the available data and conclude that HPAI is spread via trade rather than migration routes.

Many low-pathogenic forms of avian influenza circulate through the poultry community. Biosecurity precautions are the first line of defense against infection. Stay informed about events in your local area. Vaccines have been developed, but the USDA has stockpiled over forty million doses in reserve. The agency says the reserves may be used in the event of an outbreak and has kept the vaccines from use by private poultry owners.

Flock eradication remains the government's solution of choice in dealing with avian influenza. Records of your biosecurity practices can support an argument in favor of quarantine rather than forced killing of your flock in the event of an outbreak in your area.

Organizations like GRAIN, an international nongovernmental organization, champion the interests of small-flock owners through sustainable management and agricultural biodiversity.

Infectious laryngotracheitis is a herpes virus disease of chickens, pheasants, and peafowl that is very contagious, especially in confined flocks. Turkeys, ducks, and geese do not get symptoms but can spread the virus. Contaminated equipment can also spread it. Infected birds develop respiratory problems and may die. Effective vaccines exist that can be applied to the eyes and reduce the severity of the illness; however, carriers can still be infectious to other birds. Antibiotics can help the bird overcome secondary infections.

Newcastle disease, including **exotic Newcastle disease**, refers to a group of diseases having variously low, moderate, and severe virulence. Exotic Newcastle disease can infect and kill any bird. It is a highly contagious disease usually traced to birds smuggled from South America. Parrots can harbor the virus without showing symptoms.

All poultry can be affected, but chickens are the most susceptible and waterfowl the least. It usually presents as a respiratory disease but may also have neurologic symptoms, such as tremors and paralysis, and digestive symptoms, such as diarrhea. Birds catch it from each other or from contaminated equipment. Chicks should be vaccinated in locations where they are likely to be exposed to it. Supportive care may help adult birds survive. Chickens can develop immunity to the severe forms after catching the mild forms, which are used in some vaccines.

Salmonella enteritidis can infect eggs, making people who eat undercooked eggs and meat sick. Cook eggs and meat thoroughly to avoid getting sick.

Avian Diseases

Disease	Lifespan away from birds	Species Affected	Symptoms	Treatment
Avian Influenza	Months	All fowl, varies in severity	Ruffled feathers; soft-shelled eggs; depression and droopiness; sudden drop in egg production; loss of appetite; cyanosis (purplish-blue coloring) of wattles and comb; edema and swelling of head, eyelids, comb, wattles, and hocks; diarrhea; blood-tinged discharge from nostrils; loss of coordination, including loss of ability to walk and stand; pinpoint hemorrhages (most easily seen on the feet and shanks); respiratory distress; increased death losses in a flock	Supportive care; antibiotics to reduce secondary infections
Infectious Bursal Disease	Months	Chickens	Depressed; debilitated; dehydrated; watery, bloody diarrhea	Prevent through vaccination and biosecurity
Coccidiosis	Months	All fowl	Weight loss; poor growth; watery, bloody diarrhea; sick bird appearance (ruffled feathers, huddling, depression)	Amprolium or other cocciostat. Prevent with medicated feed for chicks
Duck Plague	Weeks	Waterfowl	Depression; loss of appetite; decreased egg production; nasal discharge; increased thirst; diarrhea; ataxia; tremors; hypersensitive to light; bloody discharge from vent; dark, bloody areas where ducks have rested; sudden death	Vaccinate prior to breeding season if geographically vulnerable; flock depopulation; premises decontamination
Fowl Cholera	Weeks	All fowl	Acute respiratory symptoms; neurologic symptoms; convulsions; and sudden death. Chronic infections may affect joints, eyes, throat, or other organs	Vaccination for turkeys and chickens; good management practices for prevention; antibiotics
Coryza	Hours to days	Chickens, turkeys, game birds, peafowl	Nasal discharge; facial swelling; sneezing; labored breathing; and fetid odor of secretions	Vaccines available; sulfa or other antibiotics to treat infected birds; premises disinfection may be necessary

Disease	Lifespan away from birds	Species Affected	Symptoms	Treatment
Marek's Disease	Months to years	Chickens, turkeys	Different varieties: paralysis; eye problems; general wasting; sudden death	Vaccination; acquire birds only from Marek's-free flocks
Exotic Newcastle Disease	Days to weeks	All fowl	Respiratory: sneezing, gasping for air, nasal discharge, coughing. Digestive: greenish, watery diarrhea. Nervous: depression, muscular tremors, drooping wings, twisting of head and neck, circling, complete paralysis. Also partial to complete drop in egg production; production of thin-shelled eggs; swelling of the tissues around the eyes and in the neck; sudden death; and increased death loss in a flock.	Vaccination in areas of contagion; premises disinfection; depopulation
Mycoplasmosis (MG, MS)	Hours to days	All fowl	Chronic respiratory symptoms like coughing, sneezing, and nasal discharge; red and weepy eyes; lethargy, loss of appetite, weight loss; sick bird appearance; in turkeys, the sinus under the eye becomes swollen	Triple antibiotic therapy in eyes, by mouth, and in water
Salmonellosis (Pullorum)	Weeks	All fowl	Diarrhea	Now virtually eliminated by NPIP testing program
Avian Tuberculosis	Years	All fowl	Progressive weight loss in spite of a good appetite; depression; diarrhea; increased thirst; respiratory difficulty; decreased egg production	None except depopulation.
Fowl Pox	Months	Chickens, turkeys, game birds	Poor growth; poor feed conversion; reduced egg production; wartlike growths; scabs on unfeathered body parts and/or diphtheritic (wet) membranes lining the mouth or air passages	Vaccination; no treatment

Pullorum disease, caused by *Salmonella pullorum*, and fowl typhoid, *Salmonella gallinarum*, have been largely eradicated by the National Poultry Improvement Plan. Blood tests showing that birds are not infected are required to participate in most poultry exhibition events. State and local governments work with the federal government to administer these programs and train testers. Contact your local NPIP representative.

Vaccines are available for **fowl pox**, a viral disease that occurs in two forms. One form is spread by biting insects like mosquitoes and by contamination of open cuts in the skin. The other form is spread by inhaling contaminated air. The first is a minor illness that causes sores on the comb, wattles, and beak, and most birds recover. The second causes severe respiratory symptoms, including a diphtheritic membrane in the respiratory passages. Many birds die from it.

Fowl cholera is a respiratory disease caused by the bacteria *Pasteurella multocida*. Poultry can catch it from other birds, rodents, and even from humans, who can carry the germ without being sick. Flocks at risk of infection, meaning they live in areas with a history of fowl cholera or near a current infection, can be vaccinated. The vaccination has its own risks and limitations. Live virus vaccines may have significant side effects, whereas killed virus vaccines may not be effective against the strain threatening the flock.

Aspergillosis is the most frequently seen of several fungal diseases that affect poultry. The fungus is often found in feed and litter. Mycotoxins produced by fungi like aspergillus are often invisible and tasteless. The illness is not contagious from bird to bird. They catch it from inhaling the mold organisms that are growing on their feed where they live. There is no treatment. The best tactic is prevention—keeping the living environment clean and mold-free.

INTERNAL AND EXTERNAL PARASITES

There are many kinds of lice and mites that infest birds and are likely to end up on your poultry. They can survive away from the host birds for months, to be picked up later on clothing or other birds and transferred to your home flock. They live in the crannies of wood structures. Many flock owners don't wait until they see signs or symptoms of parasite infestation but rather treat their flocks routinely once or twice a year.

Mites are blood feeders and will drain off your birds' energy and resistance to disease. Infected birds get weak and lose appetite due to anemia. Some mites come out only at night, so you may never see them.

About forty species of chewing **lice** infest fowl. They can be seen on the skin and feathers. The eggs look like white clusters at the base of the feathers. Dirty feathers, sores, and scabs, especially around the vent, may indicate infestation.

Chemical insecticides are the treatment of choice. Pyrethrin sprays or dusts are available. Carbaryl is effective, sold as Sevin powder and Purina Poultry Dusting Powder in a carbaryl/sulfur combination. Remove all litter, and treat the housing as well as the birds. Because chemical treatments don't kill the eggs, repeated treatments are needed to eliminate lice from the flock.

Poultry may be infested with lice, such as the Chicken Body Lice shown here. The females are yellowish in color and live about twelve days, laying up to four eggs each day. Egg masses may be seen as white clusters at the base of the feathers, especially around the vent. They feed on feathers and skin debris and secretions. Pyrethrin dust and other chemical insecticides eliminate them. Treat as soon as you see symptoms, as severe infestations can debilitate birds. Read labels and use insecticides carefully as they can affect use of eggs and meat. *Clemson University, USDA Cooperative Extension Slide Series*

Distinguishing between Lice and Mites

	Lice	Mites
Size	2–3 mm long	1 mm diameter (ground pepper)
Speed	Fast-moving	Slow-moving
Color	Straw-colored (light brown)	Dark reddish black
Egg location	Base of feather shaft	Along feather shaft
Egg color	White	White or off-white
Best detection time	Daytime	Nighttime or daytime
Location	Lives only on host	Lives on host and in environment

The Ohio State University Department of Veterinary Preventive Medicine, Carrie L. Pickworth Avian Disease Investigation Laboratory, and Teresa Y. Morishita, Extension Poultry Veterinarian, Ohio State University Extension

Insecticide roost paints are available to kill pests, which is especially useful for combating mites that only come out at night, hiding in cracks. Some flock owners paint wooden roosts and structures with used motor oil, but that makes the birds' feathers messy. Kerosene is also used to paint the wood birds encounter and repel parasites, but it is flammable, so use with great care.

Ticks suck blood and can cause anemia and even death in young birds. Guinea fowl happily eat ticks, so if you live in an infested area, adding them to your operation may be the answer. Otherwise, birds can be treated with pyrethrin powder or spray to repel ticks.

Birds pick up internal parasites like **worms** and **flukes** from intermediate hosts like insects, earthworms, slugs, and snails. Keep food and water clean and dry to avoid contamination by earthworms. Birds on pasture will encounter these pests. If your birds are doing poorly for no obvious reason, internal parasites are a likely culprit. Get veterinary advice for a specific diagnosis to target treatment at the organism responsible. Pathology laboratories can examine the internal organs of a dead bird to determine what has infested it. Knowing the cause can direct treatment for the rest of the flock.

Coccidiosis is caused by a protozoan parasite and can affect any poultry. The eggs are commonly found in the soil. Most chick starter feeds include a low dosage of cocciostats,

Roundworm Medications

Piperazine: Approved for chickens and turkeys

Hygromycin B: Approved for all poultry, waterfowl, and game birds

Coumaphos: Approved for all poultry, waterfowl, and game birds

Thiabendazole: Approved for pheasants

Tapeworm Medications

Butynorate: Approved for all poultry, waterfowl, and game birds usually Amprolium, which is a medication to fight cocci infection. After the first year, the birds' immune systems have encountered some cocci, and their natural immunity has developed to the point where they no longer need medication. They will be immune for the rest of their lives. Do not feed medicated starter to goslings or cygnets, which may overeat and overdose themselves.

Tapeworms and most **roundworms** infest the intestinal tract. **Flukes** can embed themselves in all parts of the bird's body. Chemical insecticides like Piperazine in feed or water and Coumaphos in feed address roundworms. Butynorate in feed kills tapeworms. It must not be given within twenty-eight days of slaughter. There is no treatment for flukes in poultry. They can be controlled by keeping water clean and keeping birds away from the intermediate hosts: snails, dragonfly larvae, and mollusks.

PREVENTING FLIGHT

All poultry except the flightless ratites can and will fly. Heavy geese and turkeys are less able to fly long distances, but their abilities to take wing over fences must be considered in housing and protecting them.

The primary wing feathers can be clipped, making flight difficult or impossible. Use strong clippers, such as tin shears. Only one wing needs to be clipped to incapacitate flight. Leave the outermost two feathers intact so that the wing retains its appearance. Clip the rest of the primaries but not the secondaries. The bird needs those to keep warm.

To continue to prevent flight, feathers will need to be clipped at least annually, after the bird has molted and grown new, complete feathers. Guinea fowl re-grow their feathers every four to six weeks. Many birds lose interest in flying after being clipped for a year.

Pinioning is a surgical procedure involving the amputation of the end section of the wing where the primary flight feathers grow. It is done on young birds, before their wings are fully developed. That section of the wing never grows back, so nothing further needs to be done to prevent the bird from flying.

The operation is more commonly done on large birds like swans, which are excellent flyers and may be under legal constraint not to leave their home. Birds that are not native are not welcome to take up residence in the

wild. Mute swans have become a problem in some places, and New Hampshire requires them to be pinioned. The state of Washington does not allow them at all.

VETERINARY ADVICE

It is to your advantage to have a relationship with a veterinarian who is willing and able to care for the kind of birds you raise before they get sick. Domestic poultry are a specialty that not all are prepared to treat. Other poultry owners in your area can recommend individuals. The Association of Avian Veterinarians provides information about its members on its website, including a page to locate one of its members in your area. *Backyard Poultry* magazine regularly publishes a list of veterinarians who are interested in including domestic poultry in their practices.

Contact veterinarians and discuss your situation with them before you have a disease outbreak crisis. Local veterinary advice can help you make decisions on vaccination and other preventive measures for your flock.

PREDATORS

Predators are the single worst nightmare for poultry keepers. Unfortunately, the predator-proof poultry coop has not yet been built. Build the strongest, most secure housing you can to foil the marauders in your area.

Raccoons and opossums are almost everywhere. Dogs allowed to run loose can become destructive, and

Opossum can be significant poultry predators, eating both birds and eggs, but they may also control rats and mice, which could be worse predators to your flock. Fence and house your birds securely. Find a balance between domestic livestock and wildlife. *Shutterstock*

Chicken wire fencing is a good choice to protect poultry from predators. Consider space, available outbuildings, climate, and local wildlife in making decisions as to the kind of shelters your birds will need. Remember that predators are persistent and will search diligently for any weak place in the fence, including digging under and going over the top. *Shutterstock*

domestic cats can take chicks and small birds like quail and Chukars. Coyotes, foxes, weasels, minks, skunks, and rats are four-legged culprits. Snakes take eggs and chicks. Snapping turtles take ducklings. Owls, hawks, and occasionally vultures attack from above. Mountain lions and bears are rare but possible predators.

All these critters will be persistent in trying to get to your birds. Protection starts with a secure perimeter. This can be either a fenced pen or fenced pasture. Concentric pens are designed to enclose the most vulnerable birds at the center, with guardian animals and birds that have more defensive resources in outer circles. Electric fencing can be part of the solution. Many predators are diligent diggers, so secure the ground level as well. Either fencing buried in the ground or a concrete floor is effective. Metal flashing protects the bottom walls of wooden structures.

Livestock guardian dogs are very effective. Certain breeds, such as Great Pyrenees, Anatolian Shepherds,

Anatolian Shepherd dogs stand 28 to 30 inches tall at the withers and weigh between 100 and 150 pounds. They have been guarding livestock for thousands of years in Eurasia. Consider using a guard dog to protect your flock against predation. *Ballester's Double H Ranch*

An effective electric fence for predator control can be constructed with either a multiple of single-strand wires or a combination of wire mesh and single-strand wires. A mesh-wire fence is more expensive to build than a fence made from single-strand wire. However, an electrified mesh-wire fence presents a greater physical barrier to predators and requires less maintenance and fewer electrified wires than a fence using only single-strand wires. *Shutterstock*

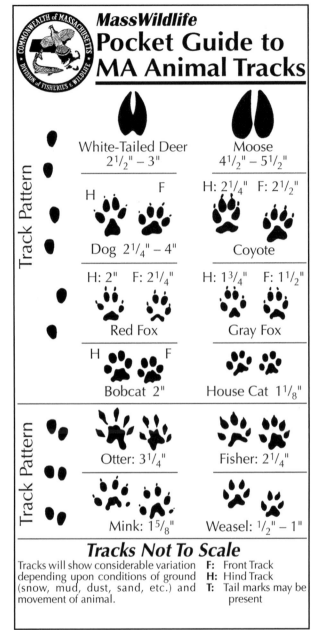

Identifying what kind of predator is attacking your birds is important to find the most effective way to resist its destruction. Tracks may be evident. Motion-sensor lights may be sufficient to convince nocturnal critters to leave your birds alone. *MassWildlife*

Akbash, and Maremma, are known for their abilities, but mixed breeds also often have the aptitude and disposition to succeed as guardians. Guard dogs must be trained to overcome their natural instincts to chase birds, and the birds must learn that the dogs will not attack them. It can be a lengthy training that requires patience as well as expertise.

Llamas and donkeys can also be effective animal guardians. Geese are capable defenders and can protect other birds. As Harvey Ussery recounted in Chapter 4, his geese protect the entire mixed flock.

Some flock keepers find keeping a radio playing helps keep predators from attacking birds. Motion sensor lights can reduce attacks.

Shooting, trapping, and poisoning will eliminate individual predators, but others are likely to move into the territory as soon as the offender is removed. These methods all run their own risks. Poison—and the poisoned carcasses—may be eaten by other animals, including your birds. Trapping and relocating wildlife is difficult and problematic. Unless they are relocated far from their home territory, they will find their way back, making your efforts pointless. If moved far enough away, relocated animals often die because they are introduced to other animals' territories and are not able to find food and shelter. Seek advice from local rangers or extension agents. You wouldn't want someone relocating their problem raccoon near you.

Merritt Clifton, editor of *Animal People*, recommends training the indigenous animals to find food by setting out eggs that for some reason would not be sold or eaten at a distance from the poultry quarters: "That way the predator learns that the place to find eggs is not inside the henhouse, but rather inside the concrete manure revetment, on the floor of the open garage or some other place."

Predation is the negative side of the human/wildlife interface. There will always be some conflict. Determine what animals are targeting birds in your area and find strategies to manage the situation with the least loss of life all around.

Mealworms

David Sullenberger, also known as Professor Chicken, is an advocate for small-flock ownership and commonsense poultry husbandry. He freely dispenses political opinions with his poultry advice. Among his suggestions is to raise your own mealworms—the larvae of small black beetles—as a tasty and economical protein supplement for turkeys, ducks, guineas, and other fowl. They are easily raised from worms purchased at a pet store or bait shop.

Equipment:
- Fine-mesh hardware cloth screening to cover buckets
- Two 5-gallon plastic buckets with lids
- Silicone sealant or glue to bond the screen to the buckets

Cut the screening in a circle a little smaller than the inside of a lid so that the screen fits easily inside the lid.

Cut out the center of the plastic lid, leaving at least an inch around the edge. Glue the screen inside the lid with the silicone adhesive. This gives you a well-ventilated cover for the bucket. Allow the silicone adhesive to cure in a well-ventilated place to dissipate the fumes. Follow product directions.

Place three inches of corn meal, bran, poultry-laying mash, flour, or other grain product in the bottom of one bucket. Spray it with clean water to dampen it, but don't soak it. Add half your mealworm livestock starter on top. Put the screen lid on and mark the date. Start the second bucket the same way three days later. Keep it in a warm place, 65 to 85 degrees Fahrenheit, isolated from human food and poultry feed. The beetles will readily infest other foods.

Worms do best in a moist, but not soggy, environment. Too much wetness can foster mites and mold. Additional moist fresh vegetables, such as carrots, potatoes, and lettuce, are appreciated. Set them on top of the mash on a thin piece of metal or wood that can be removed for cleaning.

The starter worms will grow into pupae and finally emerge as beetles, which will lay more eggs that will become the mealworms you harvest.

An egg carton makes a good feeder for ducklings. Keep water available to help the dry food go down, otherwise ducklings can choke. They typically take a mouthful of food and wash it down with water. *Art Lindgren*

FEED AND NUTRITION

Each poultry species has its own nutritional needs addressed in the individual chapters, but they all have common needs as well. They all require feed adequate to their age and numbers and appropriate to their species. All require clean fresh water. Poultry need grit to break up and digest their food because they do not have teeth. Grit in the gizzard grinds the food into digestible particles. Without grit, food passes through undigested, and birds are not able to extract full nutritional value from it.

Nutritional deficiencies manifest in various weaknesses, malformations, and even death, particularly in hatchlings. Commercially prepared feeds are required to be tagged with accurate descriptions of what they contain. Making your own feed keeps you close to knowing exactly what your birds are eating. Make sure you cover all the nutritional bases. Birds on pasture eat a varied diet, but it rarely fulfills their nutritional needs. Supplement with commercial preparations as needed.

Keep all feed clean and dry. Moisture allows mold to grow, which can be lethal. Spilled feed attracts rodents, which can carry parasites and disease.

Net Weight Bulk or 50 lbs. (22.68 kg)

SAUDER FEEDS

831-00
GAME BIRD STARTER

Mini Pellet

GUARANTEED ANALYSIS

Crude Protein, Min	26.00%
Lysine, Min	1.60%
Methionine, Min	0.45%
Crude Fat, Min	3.50%
Crude Fiber, Max	4.50%
Calcium, Min	1.10%
Calcium, Max	1.35%
Phosphorus, Min	1.00%
Salt, Min	0.20%
Salt, Max	0.50%

INGREDIENTS

Plant Protein Products, Grain Products, Porcine Meat & Bone Meal, Dicalcium Phosphate, Calcium Carbonate, Processed Grain By-Products, Molasses Products, Hydrolyzed Vegetable Fat and Animal Fat, Salt, Methionine Hydroxy Analogue, L-Lysine, Choline Chloride, Manganous Oxide, Zinc Oxide, Ferrous Sulfate, Copper Sulfate, Niacin, Vitamin E, Menadione Sodium Bisulfite Complex (Source of Vitamin K Activity), Vitamin A supplement (Stability Improved), Folic Acid, Calcium Pantothenate, Vitamin D_3 Supplement, Thiamine Mononitrate, Vitamin B 12 Supplement, Riboflavin Supplement, Ethylenediamine Dihydriodide, Pyridoxine Hydrochloride, BHT and Ethoxyquin(Preservatives).

Sauder Feeds Inc.

Manufactured by:
Box 130
Grabill, IN 46741
Phone 260-627-2196

F#61 - Mini Pellet
Rev 03/07
Pink- Gamebird

All commercial feeds are required to display a tag showing the nutritional composition of the contents. Sauder Feeds in Grabill, Indiana, caters to its customers' need for specialty feeds. This Game Bird Starter tag shows the percentage of various nutrients, such as protein and fat, and lists the ingredients. Doing business with local feed suppliers supports the network of businesses that create successful local economies. *Sauder Feeds*

Eating is an important part of the birds' day. Fresh greens and other interesting food, such as forage cakes, prevent bad habits like picking at each other.

Forage Cakes, made by Resolve Sustainable Solutions, compress a full nutritional diet into a form that keeps poultry pecking and interested in their food. Birds benefit from being able to engage in their natural foraging behavior, even if circumstances don't permit them to be outdoors. The company's motto is, "We bring free range into the coop."

Chickens and other poultry retain the instincts to scratch and peck for their food. Behavior that served them well as occupants of the jungle floor makes a mess of their feed in domestic flocks. Feed gets scratched out of the feeder, onto the ground. Further scratching mixes it with dirt and droppings. Dried dust circulates in the air. Bacteria and mold feed on the nutrients, creating a breeding ground for disease.

Commercial formulations, whether pellet or crumble, start as powdered ingredients. They pass through birds' digestive tracts faster than their digestive processes can act. Nutrients are not fully absorbed. As little as 35 percent of the feed may actually nourish the flock.

Forage Cakes were created to overcome the limitations of commercial feed for domestic flocks. They are giant granola bars, formulated from tasty ingredients like cranberry seeds, leftovers of vegetable juice processing, and nuts, mixed up with crustacean meal made from crab and shrimp, boosted with catfish meal for protein. Freshly ground spices are natural antibiotics and antihelminthics, and they

ForageCakes

make the cakes unpalatable to rodents and other vermin. The ingredients are bound together with beef gelatin.

Because less waste and cleanup are involved with Forage Cakes, they provide more usable feed for the feed dollar. "Commercial feed companies are cutting back on ingredients. Processed feed includes the cost of all that processing," says Kermit Blackwood, company founder. "It's a way to save money. People can see that they can afford to raise poultry."

Forage Cakes come in formulations for chickens, baby chicks, and mixed flocks. Hunter Kibble is formulated for zoos but can be used as a high-protein supplement for domestic flocks. The products are available through major poultry suppliers and some feed stores.

Hanging feeders set at the height of the birds' heads help reduce waste. Provide sufficient feeders for all birds to eat. Keep feed dry. Moist feed grows fungus that could kill your birds. Clean feeders frequently. *Metzer Farms*

Hatcheries support prospective small-flock owners with starter sets such as this one from Metzer Farms. It includes all the basics to greet ducklings: brooder guard to enclose them safely, feeder and waterer, vitamins, brooder lamp and bulb, and thermometer. Add ducklings, litter, and feed and you're launched. New heat-conserving technology allows hatcheries to ship as few as two ducklings or goslings through the mail. *Metzer Farms*

LEGAL CONSIDERATIONS

Many kinds of poultry and birds are legal within urban boundaries. However, each governmental unit is different. Check with your local authorities. Poultry are welcome in rural settings.

Beyond being legal, be considerate. Be a good neighbor by keeping your birds clean. Avoid unpleasant smells. Compost or dispose of manure to prevent attracting insects. Birds may be legal but can still create problems by becoming a nuisance.

In urban and suburban settings, attractive buildings and landscaping can help your birds fit in with their neighborhood. This is an example of good fences making good neighbors. Keep your birds secure. They will attract predators, and no one wants your ducks waddling across the streets. Landscaping can muffle their vocalizations. Keep feed secure to avoid attracting rodents.

Fresh eggs and friendly invitations to acquaint neighbors with your birds can go a long way. Many people know little about poultry but are delighted to learn. Offer your services as a resource to schools and local agricultural groups like 4-H and the National FFA

Organization. Make yourself and your birds an asset to the community. Getting to know a network of people can also help you in the event that your poultry are challenged. Seek help and advice from people in your area and further if a legal problem develops. Pursue solutions that benefit everyone.

Migratory waterfowl are protected by state and federal laws. Permits are required to keep wild birds legally. Domestic birds are exempt. If you are in any doubt about the status of the birds you are considering acquiring, check with your state department of fish and game, parks and wildlife, or natural resources.

SELLING MEAT, EGGS, AND OTHER PRODUCTS

Domestic birds offer a wide variety of products for which there are markets and eager customers. A small, local operation can rely on word of mouth to sell out. Larger operations require more support. Consider collaborating with other small producers through farmers' markets, cooperative agreements, and marketing associations.

Children can share responsibilities for caring for poultry. Adult supervision is required, making poultry occassions good projects for building family relationships. Agricultural organizations provide structured activities for formal learning. Children and adults spend many happy hours with their birds. *Shutterstock*

Birds love leftovers and kitchen trimmings, but such treats can be poisonous. Chocolate is toxic to birds, as are avocados.

Tobacco smoke is irritating to birds. On the other hand, tobacco in the bedding can reduce external parasites, and the small amount birds eat may have the same effect on internal ones. A pinch of tobacco is sometimes recommended to worm large birds like geese.

Common household cleaners may include chemicals like chlorine bleach and ammonia, which are toxic to birds. Rinse well after washing and allow to dry fully any housing, feed, water dishes, and other objects with which your birds will come in contact. As an alternative, soap and water or vinegar and water make effective, safe cleaning products.

New products like carpeting may release formaldehyde fumes. Air them completely before allowing your birds to have contact with them.

Paint and wood stains may have ingredients that are toxic to birds. Use with care, obey all instructions, and ascertain whether they are safe for birds. Products labeled "Safe for Pets" usually refer to dogs and cats and may be toxic to birds.

Chemical pesticides, insecticides, herbicides, and medicines like antibiotics play an important role in keeping birds healthy. They are powerful and must be used with care. Always follow all label directions. More is not better. Used appropriately, they make our birds stronger and healthier.

Ducks can be slaughtered by cutting their heads off, either on a chopping block or hanging in a killing cone. Hang the bird upside down to drain blood out completely. Suspending it also prevents the meat from being bruised. *Art Lindgren*

Extension agents offer resources for information on legal requirements for selling animal products and contacts to other producers. Collaboration with other likeminded farmers can help develop markets and relationships with wholesale and retail outlets. Support your breed organizations.

SHOWING

Chickens, ducks, geese, turkeys, and guinea fowl are recognized by the American Poultry Association. APA members earn points leading to recognition as Master Exhibitor, Grand Master Exhibitor, and Hall of Fame Exhibitor. Points are awarded for exhibiting birds that win Champion or Reserve Champion in APA-sanctioned shows. Number of points is proportional to the size of the show, and the total number of birds exhibited.

Other birds are often shown at poultry shows, too, and judges are often happy to examine them. Local poultry clubs may have specialty categories. Game birds may have their own shows. They are usually welcome at the sale table. Shows are an excellent place to acquire new birds and sell your surplus.

Shows attract other breeders and aspiring breeders. Attending shows and participating in them gets you and your birds out into the community of shared interests. Mingling with other breeders creates relationships for sharing information and trading birds. Senior breeders can give novices the benefit of their years of experience.

Shows introduce the public to poultry. This is where youngsters and their parents learn how they can get

Poultry shows bring hundreds or even thousands of birds and their owners together. Take the opportunity to meet others who are interested in your breed. Get a close look at other breeds. Bring surplus stock to sell or exchange. Make new friends and renew acquaintances. Prepare your birds well by taming them and getting them accustomed to cages and judging sticks with practice and reward. Shampoo them if needed before the show. Be a gracious winner and an appreciative loser. Learn from judges and other exhibitors to improve. Never dye feathers or otherwise misrepresent your birds. *Courtesy of author*

Vendors and poultry organizations attend shows. Feed and equipment representatives have their products on display for you to examine and get answers to your questions. Poultry organizations such as the Society for the Preservation of Poultry Antiquities present their information. Poultry publications display their books and magazines. Shows bring all the information into one place for poultry enthusiasts. *Courtesy of author*

involved. The more the public knows and understands about small-flock raising, the more community support the endeavor will have.

Shows are a lot of fun, with many events beyond the exhibition. People whose paths might otherwise not cross get together in a cheerful atmosphere. The *Poultry Press* is the major publication for exhibitions.

NATIONAL ANIMAL IDENTIFICATION SYSTEM

The National Animal Identification System (NAIS) is a federally operated livestock ID system intended to trace animal movements in the event of a disease outbreak. It remains controversial. The first phase of the three-phase program is premises identification. Although officially voluntary at the national level, several states have already made it effectively mandatory by not allowing farmers to sell products from premises

that are not NAIS-registered. Some states sparked controversy when they required premises registration for participation in 4-H and FFA events.

The program entered its second phase, individual animal identification, in 2008. You may be required to identify each of your birds with a leg or wing band. Check local requirements.

The third phase will require all animal movements to be reported within forty-eight hours. Movements include sales and shows—anything that takes livestock off or brings it to your farm. Keep informed as to the progress of NAIS in your state.

Some farmers oppose NAIS, on religious, philosophic, and practical grounds. They consider individually tagging every bird labor intensive. Costs of tags and tracking devices add to operations. Small-flock keepers resist reporting every movement of birds on and off the farm as an onerous chore.

CHAPTER 11

. . . .

CONCLUSION

Keeping domestic poultry has been almost a cultural universal since long before people wrote about it. Fascination and delight with birds and their willingness to live with us have kept poultry and humans together through millennia. The future should continue to bless us with the enjoyment and bounty of small flocks of birds.

Major challenges face small-flock owners. The regulations required by NAIS may prove too onerous for small-flock owners. Highly pathogenic avian influenza, or the measures taken against it, may eliminate entire flocks and even wipe out rare breeds entirely.

High oil prices are recalibrating costs of feed and fuel. How these economic changes will play out for small-flock owners remains to be seen. The increase in markets for local food may well favor the small producer over the industrial food corporations.

Increased emphasis on "staycations," vacation time spent close to home, and agricultural tourism offer additional opportunities. Farm tours, day camps, and you-pick produce all draw in paying customers. Chicks naturally attract visitors in the spring, providing an opportunity to connect farmers with interested novices and help them get started. Poultry shows provide networking opportunities. Cooperative marketing organizations can increase marketing support among operations that offer varied products and services, such as eggs, feathers, meat, and hide.

People have many reasons for keeping and breeding domestic birds. The variety is so wide that any interested

Livestock graze at a small dairy farm in western Maryland. The U.S. Department of Agriculture defines "small farms" as those averaging $50,000 in gross sales annually—which net, on average, around $23,159. Increased consumer interest is driving markets in grass-fed beef and pasture-raised pork as well as pastured poultry. Bestselling books such as *The Omnivore's Dilemma* **by Michael Pollan and** *Animal, Vegetable, Mineral: A Year of Food Life* **by Barbara Kingsolver reached a wide range of readers.** *Scott Bauer*

fancier can find a breed that suits his or her personality and life circumstances. City dwellers can keep pigeons, and some are keeping chickens.

Breeding birds can be financially rewarding, even a full-time job, if that's what you would like. Business plans for small farms and small-flock poultry operations, such as Ostrich.com's materials on starting an ostrich farm, and sample budgets can help determine whether this is right for you and what it will take. Help is available.

Tying your farm promotions to seasonal celebrations attracts customers. Relating products to customers' lives helps them relate their household needs to small producers. Agricultural operations give children valuable knowledge about food and culture as well as offering wholesome family activities. *Shutterstock*

Farmers' markets are a great place to sell poultry products. They also give small-flock owners contacts within the sustainable agriculture community. Cooperative marketing helps spread the word. Offer printed brochures to help customers understand your operation, its significance to food production and value to them. *Shutterstock*

Events such as the annual Tour d'Coop in Raleigh, North Carolina, attract increasing numbers of visitors. Proud coop owners enjoy sharing the solutions they found to their individual situations. Prospective owners learn how they can manage a small flock. The Tour d'Coop raises money for charitable causes. Bringing people with common interests together creates the community that supports healthy lifestyles. *Rick Bennett*

Government agencies have lots of information and dedicated personnel. If they aren't familiar with the birds you want to pursue, become their resource. Be the change you are seeking.

People have looked at birds in wonder and delight, astonished at their aerial acrobatics and inspired by their family devotion. The beauty of their feathers has tempted humans to kill the birds in order to array themselves in their finery. Fanciers and breeders find joy in these birds. Careful husbandry can ensure that generations yet to come will continue to enjoy them.

This Dorking hen lives at Plimouth Plantation in Massachusetts. She is part of the living history collection of historic livestock breeds that includes Kerry cattle, Milking Devon cattle, San Clemente and Arapawa Island goats, and Tamworth swine. Live collections play an important role in raising public awareness of traditional breeds as well as focusing attention on breeding operations. *Courtesy of author*

The National Poultry Museum is housed at the National Agriculture Center and Hall of Fame in Bonners Springs, Kansas. This artist's drawing is a projection of a building that some day might house the full collection. Loyl Stromberg of Minnesota has been a leader in establishing the first building and donating equipment and materials from his personal collection, as have others such as John Skinner of Wisconsin.

Agricultural Tourism Glossary

Agricultural tourism: Refers to the act of visiting a working farm or any agricultural, horticultural, or agribusiness operation for the purpose of enjoyment, education, or active involvement in the activities of the farm or operation.

Certified farmers' market (CFM): A location approved by the county agricultural commissioner, where certified farmers offer for sale only those certified agricultural products they grow themselves. Other agricultural and non-agricultural products may be sold at the markets, depending on regulations and market rules.

Community-supported agriculture (CSA): A partnership between consumers and farmers in which consumers pay for farm products in advance and farmers commit to supplying sufficient quantity, quality, and variety of products. This type of arrangement can be initiated by the farmer (farmer directed) or by a group of consumers (participatory).

Direct marketing: Any marketing method whereby farmers sell their products directly to consumers. Examples include roadside stands, farm stands, U-pick operations, community-supported agriculture or subscription farming, and farmers' markets.

Farm visits or **stays:** The activity of visiting a farm for short periods of time or overnight stays for the purpose of participating in or enjoying farm activities and/or other attractions offered.

Rent-a-tree operations: These are arrangements where customers "rent" or "lease" trees from farmers. The consumer pays the farmer at the beginning of the season, the farmer takes care of the trees on the farm, and either the farmer or the customer does the harvesting.

Roadside stands: Also known as farm stands, refers to any activity where the farmer sells agricultural and value-added products from his farm directly to consumers at a stand or kiosk located on or near his farm or along a road near the farm.

Rural tourism: Recreational experience involving visits to rural settings or rural environments for the purpose of participating in or experiencing activities, events, or attractions not readily available in urbanized areas. These are not necessarily agricultural in nature.

U-pick or **pick-your-own operations:** These are fruits and farms or orchards where the customers themselves harvest the fruits or products, such as a pick-your-own strawberry patch. The prices the customer pays for the volume harvested will usually be higher than what the grower would get from a broker.

Value-added: Any activity or process that allows farmers to retain ownership and that alters the original agricultural product or commodity for the purpose of gaining a marketing advantage. Value-added may include bagging, packaging, bundling, pre-cutting, and so on.

Ramiro Lobo, Farm Advisor
UC Cooperative Extension, San Diego County

Swans have become the identifying symbol of Lakeland, Florida, the City of Swans. The city commissioned one hundred artists to create swan sculptures to raise money for its swan program and other nonprofit organizations in 2002. The sculptures were sold at auction and are displayed around the city. *W. Thomas Miles*

A LOOK TO THE FUTURE

Fowl trust sites have long been my dream and the Society for Preservation of Poultry Antiquities' goal. Because poultry vary so much and are often regionally or even locally distinct, more than one would serve a perfect world. Immigrants brought their own preferred breeds and kept them separate or introduced them into local flocks, and the history of poultry reflects the unfolding of human history.

The sites would serve as conservators of historic breeds and laboratories for poultry research. Poultry breeding is an art as well as a science. The genetic diversity of these historic breeds includes natural disease resistance and adaptive behaviors. Flocks of rare breeds would be kept, both to supply new and existing owners with stock and

Small-flock owners often develop warm companion relationships with their birds. Those hand-raised from hatching will remain friendly. Older birds may be wary but are usually willing to extend a beak of welcome to an outstretched hand bearing food. Chickens and turkeys have been used as therapy animals in nursing homes and as interpretive animals at living history museums. Their value to our lives is limited only by our imagination. *Shutterstock*

to serve as educational facilities to the general public. Birds that are naturally tolerant or resistant to disease offer a clear advantage. Flocks would be bred for genetic research. Natural immunities would be explored for advantages over artificial immunities. "If breeds become extinct before their disease resistance qualities have been identified, genetic resources which could greatly contribute to improving animal health and productivity are lost forever," concludes the United Nations Food and Agriculture Organization, in its *State of the World's Animal Genetic Resources 2007* report.

The sites would maintain libraries of academic research, popular knowledge, and rare books. Historic literature not only documents events of the past but often contains knowledge that has been overlooked in the rush to reach higher feed-conversion ratios for commercial slaughter. Studying methods of the past can lead to insights for future poultry management.

Artists would work with the birds as subjects. Poultry art and photography could attract more and varied artists to portray the birds that have fascinated us for millennia. Studios could be provided and accommodation made to artists. Television programs would originate from the sites, documenting the breeds and making those records

Products such as this counted cross-stitch pattern kit (*right*) and quilt design (*above*) encourage creative projects that involve traditional breeds. Stitchery handwork is a restorative as well as an artistic occupation. The finished project can inspire others to learn more about traditional breeds and express their own artistic side. *The Posy Collection, Critter Pattern Works*

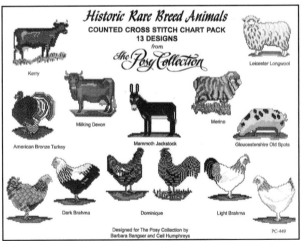

available to the public. Regular programming could offer advice to breeders as well as general-interest videos to the public.

Classrooms and lesson plans would help serve and benefit visiting school groups. Interpretive chickens, turkeys, and other willing birds could acquaint children with the feathered world. Programs would allow young people to take regular roles in bird husbandry, exposing more children to poultry keeping. Older students would be welcomed into internships. Their help could be compensated not only financially but with breeding stock to set up their own flocks. University students and professors would be welcome to use the sites' resources for their research.

Lee Becker is one of the contemporary artists taking an interest in traditional breeds. This oil painting of Barred Rocks is part of her Rare Breeds series, on display at the Lindsborg, Kansas, Arts Council. Poultry have always captured human attention. Their beauty and unique plumage colors and patterns are among the driving influences to develop and perfect breeds. *Lee Becker*

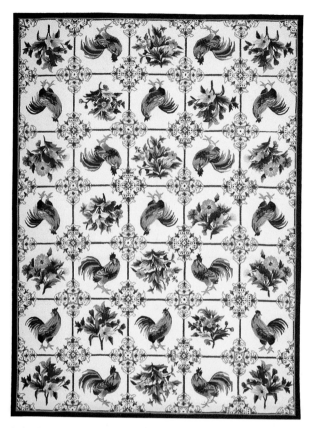

Colorful plumage of traditional chickens, reminiscent of Old English Games, is a popular decorative motif. This rug, in Nourison Rugs' Country Heritage Collection, is available as a small throw rug or a large 8×11. *Nourison Rug*

Culls from the flocks could be offered to local food outlets. Historic vegetables and fruits would be raised, integrating the birds into the operation, such as using geese to weed the fields. Sustainable methods would be used to create a Model Garden. Produce and meat and eggs from the birds would be offered in the site's restaurants. Permanent chefs would offer cooking classes and train interns.

Sites might organize birds and food around major geographic groupings: The Asia section could feature Silkies, Cochins, Phoenix, India Runner, and Mandarin ducks, and China geese. The restaurant in that area would serve roast duck garnished with bok choy and peppers, grown in the site's gardens. Cotton Patch geese would be employed to clear grass from the gardens. Turkeys and guineas would be sent in to eat insect pests. The North American section would

feature turkeys, Muscovy ducks, and Dominique and Java chickens, with the restaurant serving fried chicken with mashed potatoes and gravy. The possibilities are limited only by the imagination.

Classes in poultry husbandry, live plucking, traditional breeding methods, judge training, and other subjects would be offered. The sites would respond to community needs by sharing resources. Teachers would be able to arrange a field trip or a classroom visit from an interpreter accompanied by a turkey, several chickens, and a duck. Therapy chickens would visit hospitals and nursing homes.

Build it and they will come, is the folk wisdom. Modern nature tourism demonstrates that it can be true. The Lady Bird Johnson Wildflower Center in Texas attracts seventy thousand visitors annually to learn about native plants. Imagine what poultry would bring!

REFERENCES

American Livestock Breeds Conservancy. *How to Raise Heritage Turkeys on Pasture*. Pittsboro, NC: American Livestock Breeds Conservancy, 2007.

American Poultry Association. *American Standard of Perfection*. Mendon, MA: American Poultry Association, 1998.

Andrews, Ted. *Animal-Speak: The Spiritual and Magical Powers of Creatures Great and Small*. St. Paul, MN: Llewellyn, 1998.

This Buckeye rooster shows his rich chestnut plumage. Buckeyes are the only American breed whose development is credited to a woman, Mrs. Nettie Metcalf. She perfected the breed, and it was recognized by the *Standard* in 1904. The APA *Standard of Perfection* and the ABA's *Standard* are the required references for exhibiting and judging poultry. *Bryan K. Oliver*

White Dorking chickens were included in the first APA *Standard* in 1874. Dorkings, like other traditional breeds, should be given time—as much as a year—to mature before culling poor specimens. They grow more slowly than industrial hybrid breeds. *Robert Gibson*

Appleyard, Reginald. *Ducks: Breeding, Rearing and Management*. London: Poultry World Limited, 1949.

———. *Geese: Breeding, Rearing and General Management*. London: Poultry World Limited, 1948.

Ashton, Chris. *Domestic Geese*. Mid Glamorgan, England: WBC Book Manufacturers, 1999.

American Bantam Association. *Bantam Standard*. Augusta, NJ: American Bantam Association, 2006.

Beebe, William. *A Monograph of the Pheasants*. London: Witherby, 1918–1922.

Bender, Marjorie, D. Phillip Sponenberg, and Donald Bixby. *Taking Stock of Waterfowl: The results of the American Livestock Breeds Conservancy's Domestic Duck and Goose Census*. Pittsboro, NC: American Livestock Breeds Conservancy, 2000.

Bertram, Brian C. R. *The Ostrich Communal Nesting System (Monographs in Behavior and Ecology)*. Princeton, NJ: Princeton University Press, 1992.

Blechman, Andrew. *Pigeons: The Fascinating Saga of the World's Most Revered and Reviled Bird*. St. Lucia, Australia: University of Queensland Press, 2007.

Brown, A. F. Anderson, and G. E. S. Robbins. *New Incubation Book*. Blaine, WA: Hancock House, 2002.

Brown, D. *A. Guide to Pigeons, Doves & Quail, Their Management, Care & Breeding*. New South Wales, Australia: Australian Birdkeeper, 1995.

Burnham, Geo. P. *The History of the Hen Fever: A Humorous Record*. 1855. Ann Arbor: University of Michigan Library, 2005.

Chatterton, F. J. S. *Ducks and Geese and How to Keep Them*. Rev. ed. London: Cassell, 1951.

Cracraft, J. "Continental Drift, Paleoclimatology and the Evolution and Biogeography of Birds." *Journal of Zoology of London* 169, 455–545.

Damerow, Gail. *The Chicken Health Handbook*. North Adams, MA: Storey, 1994.

Eastman, Maxine. *The Life of the Emu*. Sydney, Australia: Halstead Press, 1969.

Feduccia, Alan. *The Origin and Evolution of Birds*. New Haven, CT: Yale University Press, 1996.

Ferguson, Jeannette. *Gardening with Guineas*. Waynesville, OH: FFE Media, 1999.

Gage, Laurie J., DVM, and Rebecca S. Duerr, DVM. *Hand-Rearing Birds*. Ames, IA: Blackwell, 2007.

Gardner, Geoffrey, Fanchon Funk, Sheila Bolin, Rebecca Webb Wilson, and Shirley Bolin. *Swan Keeper's Handbook*. Malabar, FL: Krieger, 2003.

Check out the history of Runner ducks at a local university library. Libraries have cooperative arrangements with other libraries, so nearly any book can be made available. *Bryan K. Oliver*

Gough, G. A., and J. R. Sauer. *Patuxent Bird Glossary.* Laurel, MD: Patuxent Wildlife Research Center, 1997.

Green-Armytage, Stephen. *Extraordinary Chickens.* New York: Harry Abrams, 2000.

———. *Extra-Extraordinary Chickens.* New York: Harry Abrams, 2005.

———. *Extraordinary Pheasants.* New York: Harry Abrams, 2002.

———. *Extraordinary Pigeons.* New York: Harry Abrams, 2003.

Hastings Belshaw, R. H. *Guinea Fowl of the World.* Hampshire, England: Nimrod Book Services, 1985.

Heuser, G. F. *Feeding Poultry: The Classic Guide to Poultry Nutrition for Chickens, Turkeys, Ducks, Geese, Gamebirds, and Pigeons.* New York: John Wiley, 1955.

Holderread, Dave. *The Book of Geese: A Complete Guide to Raising the Home Flock.* Corvallis, OR: Hen House Publications, 1981.

———. *Raising the Home Duck Flock.* North Adams, MA: Storey, 1978.

———. *Storey's Guide to Raising Ducks.* 2nd ed. North Adams, MA: Storey, 2001.

Holthaus, Gary. *From the Farm to the Table: What All Americans Need to Know about Agriculture.* Lexington: University Press of Kentucky, 2006.

Humane Society of the United States. *Wild Neighbors: The Humane Approach to Living with Wildlife.* Golden, CO: Fulcrum, 1997.

Hyams, Edward. *Animals in the Service of Man: 10,000 years of Domestication.* Washington, DC: Cambridge-Dent Publishing, 1972.

Irvine, Lyn. *Field with Geese: A Book about the Domestic Goose.* New York: William Morrow, 1961.

Landau, Diana, and Shelley Stump. *Living With Wildlife: How to Enjoy, Cope With, and Protect North America's Wild Creatures Around Your Home and Theirs.* San Francisco: Sierra Club Books, 1994.

Buff turkeys are a rare historic variety being nurtured back to vigor by dedicated breeders. These birds have a tendency to develop the defect angel wing, as seen in the turkey at front. The group shown here includes both *toms*, male turkeys, and *hens*, female turkeys. Various types of poultry have specific names for males and females. *George McLaughlin*

Lee, Andy, and Pat Foreman. *Chicken Tractor: The Permaculture Guide to Happy Hens and Healthy Soil.* Buena Vista, VA: Good Earth Publications, 1998.

Leveille, Jean. *Birds in Love: The Secret Courting and Mating Rituals of Extraordinary Birds.* St. Paul, MN: Voyageur Press, 2007.

Levi, Wendell. *Encyclopedia of Pigeon Breeds.* Sumter, SC: Levi Publishing, 1965.

———. *The Pigeon.* Sumter, SC: Levi Publishing, 1977.

Minaar, Phillip and Maria. *Emu Farmer's Handbook, Vol. 1.* Blaine, WA: Hancock House, 1993.

———. *Emu Farmer's Handbook, Vol. 2.* Blaine, WA: Hancock House, 1998.

Myrick, Herbert. *Turkeys and How to Raise Them: A treatise on the natural history and origin of the name of turkeys; the various breeds, and the best methods to insure success in the business of Turkey growing. With Essays from practical poultry growers in different parts of the United States and Canada.* New York: Orange Judd Company, 1899.

National Research Council. *Nutrient Requirements of Poultry.* 9th ed. Washington, DC: National Academy Press, 1994.

A Dominique hen is settled on her clutch of eggs. Dominiques are a historic breed that retains the instincts to hatch eggs and raise chicks. *Bryan K. Oliver*

Oberholtzer, Lydia, Catherine Greene, and Enrique Lopez. *Organic Poultry and Eggs Capture High Price Premiums and Growing Share of Specialty Markets.* Washington, DC: U.S. Department of Agriculture, 2006.

Pew Commission on Industrial Farm Animal Production. "Putting Meat on the Table: Industrial Farm Animal Production in America." 2008. www.ncifap.org.

Price, A. Lindsay. *Swans of the World: In Nature, History, Myth and Art.* Tulsa, OK: Council Oak Books, 1995.

Salatin, Joel. *Pastured Poultry Profit$.* White River Junction, VT: Chelsea Green Publishing, 1996.

Schatz, Sherrie, and Sheree Lewis. "Emu Oil." *Emu Today and Tomorrow,* 1996.

Schorger, A. W. *The Wild Turkey: Its History and Domestication.* Norman: University of Oklahoma Press, 1966.

Scott, George R. *The Art of Faking Exhibition Poultry: An Examination of the Faker's Methods and Processes with Some Observations on their Detection.* 1934. Reprint, Home Farm Books, 2006.

Scrivener, David. *Exhibition Poultry Keeping.* Marlborough, England: The Crowood Press Ltd., 2005.

Shaw, William T. *The China or Denny Pheasant in Oregon.* Philadelphia and London: J.B. Lippincott Company, 1908.

Siegel, Marc. *Bird Flu: Everything You Need to Know about the Next Pandemic.* Hoboken, NJ: John Wiley & Sons, 2006.

Small Farm Center, University of California. *Agritourism and Nature Tourism in California.* Davis: Small Farm Center, University of California, 2006.

Vileisis, Ann. *Kitchen Literacy.* Washington, DC: Island Press, 2008.

APPENDIX 2

• • • • • • • • • • • • • • • • • • •

GLOSSARY

atricial: Hatched naked and helpless.

balut (Filipino) or **hot vit long** (Vietnamese): A duck egg incubated for sixteen days, to be boiled and eaten. It is considered a delicacy and highly nutritious.

bank: A group of swans on the ground.

bantam: A naturally small chicken or duck.

beak: Mandibles that are sharp and pointed, compared to a bill, which is blunt and specifically refers to waterfowl.

bevy: A group of birds, especially ducks and quail.

bill: The blunt mandibles of waterfowl. The bill has a bean at the tip, which may be a different color from the rest of the bill. The bill is lined with lamellae, tooth-like serrations that strain water for food items.

breed: A line of birds that breeds true, that is, offspring that consistently resemble their parents with regard to specifics of conformation, weight, feather color, wattles, beak, skin, leg and foot color, and other characteristics.

breeding pen: A separate enclosure for adult birds selected for breeding.

carriage: The manner and style with which a bird deports itself. Varies from low (nearly horizontal) to erect (almost vertical).

caruncles: Warty fold of skin around the head of some breeds, especially Muscovy ducks.

Poultry magazines of the early twentieth century reflected the intense interest in both production and exhibition breeds. A *breed* is defined as a line of birds that breed true and resemble their parents. This advertisement appeared on the back of the *American Poultry Advocate* of August 1912.

clear: A term used in candling eggs to see if they contain a growing embryo. A clear egg is one where light easily shows through when candled. Indicates a fresh egg, or one in which no embryo is growing.

cob: A male swan.

crest: A showy tuft on the head.

crossbreed: The offspring of two different breeds. Breeds within a species are sufficiently similar to interbreed easily, but their offspring may be unpredictable and will not breed true. Crossbreeding over generations is a way to produce new breeds.

crown: Top of head.

dewlap: The fold of skin that hangs under the heads of some geese.

A *crossbreed* is defined as the offspring of two different breeds. They do not breed true but can be a way to add desirable traits to established breeds. Runner ducks have been crossed with many other breeds, to capture their egg-laying ability. They are recognized in eight color varieties, including Fawn and White as shown here, and raised in many more unrecognized colors. Breeders continue to develop more color varieties of this useful and historic breed. *Metzer Farms*

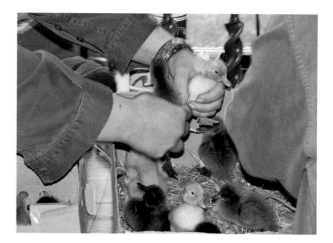

Hatchling is the word for any newly hatched bird, such as these day-old ducklings. Check your birds as soon as they arrive. Day-old birds can be safely shipped because they do not require food or water for 24 to 48 hours. These Khaki Campbell and Pekin ducklings arrive safe and sound at their destination. *Art Lindgren*

drake: A male duck. In domestic ducks descended from wild Mallards, the drake feather is a group of curled feathers at the tail.

dump nesting: When a hen persists in laying eggs in a nest after a clutch has been laid, even after another hen is incubating the clutch.

eye stripe or **streak:** Contrasting markings on the head, from the base of the bill to above and below the eyes toward the back of the head.

flock: A group of birds, usually referring to a single breed. Different breeds in one group are often called a mixed flock.

gaggle: A group of geese when not flying. In flight, they are a flock.

gander: A male goose.

gobbler: A tom turkey.

gosling: A goose chick.

gozzard or **gosherd:** One who takes care of geese.

hatchling: Any newly hatched bird.

heterosis: The vigor that sometimes results from crossing two species to produce a hybrid.

hybrid: The offspring that result from crossing two species that would not usually breed and produce viable and fertile offspring. Most cross-species hybrids are infertile. Crossing breeds within a species may result in hybrid vigor, as in the Cornish/Rock cross, but the second-generation offspring will not uniformly resemble their parents.

Plumage is a word for all the feathers on a duck, such as the white plumage on these Pekins. Three-year-old Harrison Lindgren and the ducks enjoy playing with water. Ducks are generally easygoing around children but should always have an escape route to take when they decide the fun is over. *Art Lindgren*

imprinting: The rapid learning process that establishes a behavior pattern of recognition and attraction to the first large moving object a duckling or gosling sees in its first hours. The process is inflexible, and the object on which the bird has imprinted will remain with it throughout its life.

keel: In geese, the loose fold of skin that hangs down in front of the legs. Also, the curved cartilage, which ossifies to bone, connecting a bird's breast muscles at the tip of the breastbone.

knob: The horny protuberance at the end of the bill between the eyes of African and Chinese geese.

lobe: In geese, the fold of skin that hangs from the abdomen, behind the keel. Some breeds have one lobe, and others have two.

nail: The tip of a swan's bill.

nye: A group of pheasants on the ground.

pair: A male and female of one breed. Birds that form pair bonds are called mated pairs.

party: An old term for a group of birds.

pen: A female swan.

plumage: The collective feathers covering the entire bird.

preening: Cleaning and oiling feathers with the beak or bill.

rafter: A group of turkeys.

roach-backed: Curved like a roach (i.e., convex instead of concave). Often found with wry tail. It is a serious defect that should never be perpetuated in breeding.

Roman nose: The opposite of scoop bill, a Roman nose is a convex curve of the bill. Desirable in some breeds and produces a stronger, more wedge-shaped bill.

scoop or **dished bill**: A concave depression in a duck, goose, or swan's bill, making the tip turn upward. Results in disqualification at exhibition.

skein: A flock of geese or similar birds in flight.

snood: The fleshy appendage that hangs down over a tom turkey's beak.

Standard-**bred**: Birds bred to conform to the published *Standard of Perfection* of the American Poultry Association and the American Bantam Association.

tail covert: In a bird's tail, each of the smaller feathers covering the bases of the main feathers.

trio: Two females and one male of a breed. Trios are both exhibited and bred. The trio is the basic unit of breeding, providing sufficient genetic diversity from two mothers to begin a breeding program.

twisted wing: A serious deformity in which the primary flight feathers do not fold properly next to the body but stick out at angles. This is a disqualification that should never be perpetuated in breeding.

wedge: A group of swans in flight.

wry tail: A serious deformity in which the tail is permanently displaced from the center of the body. This is a disqualification that should never be perpetuated in breeding.

FEATHER TERMS

angel wing (also **slipped wing, crooked wing, airplane wing,** or **drooped wing**): A malformation of the last joint of the wing, twisting it so that the feathers stick out. It is a disqualification in showing.

down: The fine feathers next to the skin of waterfowl. Down keeps waterfowl warm in cold and wet conditions by trapping air next to the skin. Its loft is the ability to fluff up and retain warmth.

eclipse molt: The female-coloration feathers males acquire between breeding season and winter.

frizzle: Curled feathers, found on Sebastopol geese.

fulvous: Of a dull, brownish yellow.

herl: Long stringlike material on peacock eye and sword feathers. Used to tie wet-fly patterns for fly fishing.

loft: The volume of space occupied by down in a garment or quilt, measured as thickness. A measure of the insulating ability and, thus, warmth of the product. Also, a structure for housing pigeons.

lores: The space between the eye and the beak.

loosely or **tightly feathered:** Terms describing how close the feathers are to the body. Sebastopol geese are loosely feathered compared to Egyptian geese, which are tightly feathered.

loral axe: The distinctive yellow crescent or "war-stripe" on each side of the double "striped" head of the Green peafowl.

molt: The process of replacing old feathers with new. Most poultry species change feathers once annually, but some change twice.

nuptial plumage: The feathering acquired by drakes for the breeding season.

pinfeathers: The hard sheath in which feathers grow that needs to be removed for table preparation.

pinioning: Clipping the tip portion of a baby bird's wing to prevent it from ever being able to fly.

primary feathers: The long flight feathers growing from the pinion, or outer segment, of the wing. Primaries fold under the secondaries and are not visible when the bird is not in flight. They can be clipped to prevent birds from flying, but new feathers will grow in each year and must be clipped again.

retrices: The main peacock tail feathers.

saddle: The rear part of the back extending to the tail.

secondary feathers: The long stiff feathers growing from the middle wing segments after the primaries. The secondaries fold over the primaries, exposing a triangular area called the wing bay.

sickle: The feathers on a peacock's train when raised in display, resembling a sword in shape.

smalt blue: Deep blue on peacocks.

train: Display feathers of the male consisting of sickle, fan, and ocelli feathers of varying lengths.

vermiculated: Wormlike winding or barring.

RESOURCES

ANIMAL WELFARE

The Animal Compassion Foundation.™ Whole Foods Market's nonprofit dedicated to improving the lives of farm animals. http://www.animalcompassionfoundation.org.

Animal Welfare Institute. A nonprofit organization to reduce animal pain and suffering. http://www.awionline.org.

Compassion in World Farming, A British organization founded by a farmer horrified by the development of modern, intensive factory farming. The organization campaigns peacefully to end all cruel factory farming practices. It works only on farm animal welfare. http://www.ciwf.org.uk.

Farmed Animal Net. A collaborative project of eight nonprofit organizations developed as an objective, trustworthy source of academic and industry information. http://www.farmedanimal.net.

Farm Sanctuary, Founded to combat the abuses of factory farming and to encourage a new awareness and understanding about farm animals. http://www.farmsanctuary.org.

Ducks are curious, an advantage to them as foragers. This Khaki Campbell female finds a drink in the watering can. Many resources are available to curious poultry owners. *Art Lindgren*

United Egg Producers. Welfare guidelines for the egg industry. http://www.uepcertified.com.

Viva! USA. An organization advocating for farm animals and encouraging people to adopt a vegan diet. http://www.vivausa.org.

ART

Critter Pattern Works. Quilt patterns and kits featuring poultry. www.critterpat.com

Lindsborg Arts Council. Artist Lee Becker paints portraits of historic livestock. www.lindsborgarts.org.

Nourison Rug. Imports 100 percent wool rugs, for sale through local dealers. www.nourison.com.

Posy Collection. Features cross stitch and needlework of historic Americana, including rare poultry breeds. www.posycollection.com.

AVIAN INFLUENZA

BirdLife International. A global partnership of conservation organizations working to conserve birds and achieve biodiversity; also part of a task force on avian influenza. www.birdlife.org.

GRAIN. An international nongovernmental organiztion that promotes the sustainable management and use of agricultural biodiversity based on control over genetic resources and local knowledge. www.grain.org.

U.S. Geological Survey National Wildlife Health Center. Advances wildlife and ecosystem health for a better tomorrow and provides information about avian influenza. www.nwhc.usgs.gov.

CULTURE AND SOCIETY

Agricultural tourism. The University of California Small Farm Program provides supporting materials. www.sfc.ucdavis.edu/agritourism.

Birdwatching survey. The U.S. Fish & Wildlife Service provides supporting materials. http://library.fws.gov/nat_survey2001_birding.pdf.

Cornell Lab of Ornithology. The Macaulay Library is the world's largest online archive of animal and bird sounds. www.birds.cornell.edu/macaulaylibrary.

Katrin Becker of Mink Hollow Farm in Alberta, Canada, has an excellent list of children's storybooks as well as poultry reference books. www.minkhollow.ca/HatchingProgram/Teach/BOOKLIST.htm.

Modern Homestead. Dedicated to the skills and philosophy for food independence and more self-reliant living. www.themodernhomestead.us.

HEALTH

Cornell University College of Veterinary Medicine, Duck Research Laboratory. Offers a wealth of information about duck care. www.duckhealth.com.

Kansas State University Research and Extension Library. Provides publications about livestock health, including a helpful brochure on eliminating mites in poultry flocks. www.oznet.ksu.edu/library.

Mississippi State University Extension Service. Offers information on poultry needs, including internal parasites. www.msucares.com.

NoNAIS. Fights to protect traditional rights on farms and opposes the National Animal Identification System. www.nonais.org.

North Carolina State University Cooperative Extension Service. The poultry science department offers excellent resources, including information on internal parasites. www.ces.ncsu.edu/depts.

Professional Pest Control Products. Provides professional pest control supplies and information, including a page on ticks. www.pestproducts.com.

United States Department of Agriculture, Animal and Plant Health Inspection Service. Lists legal requirements for reportable diseases and provides information on the National Poultry Improvement Plan. www.aphis.usda.gov.

Chickens evoke the ambience of country life and rural beauty. Nourison Rug capitalizes on their appeal in this wool rug. The design is included in its Country Heritage Collection. *Nourison Rug*

Welsh Harlequin Ducks are not yet recognized by the APA for exhibition. Although they remain few in number, they are good utility birds for small-flock owners, producing as many as three hundred eggs a year. This breed, developed in the mid-twentieth century, makes a welcome addition to a sustainable agriculture operation, such as these at Yellow House Farm in New Hampshire. Stock is available from Holderread Waterfowl Farm and Preservation Center in Corvallis, Oregon. *Robert Gibson*

It doesn't take much water to entertain ducks. These Pekins and Khaki Campbells make good use of a plastic container. Water drowns lice and mites, helping reduce pests. Change water and refill as needed. *Art Lindgren*

University of Florida Institute of Food and Agricultural Sciences Extension. Provides information on safe poultry feed storage. http://edis.ifas.ufl.edu.

University of Minnesota Department of Animal Sciences. Compiles information from state extension services on local poultry conditions and maladies around the country. www.ansci.umn.edu.

LEGAL ASPECTS

Convention on International Trade in Endangered Species of Wild Fauna and Flora. CITES is an international agreement between governments. Its aim is to ensure that international trade in specimens of wild animals and plants does not threaten their survival. www.cites.org.

International Union for the Conservation of Nature and Natural Resources. The world's oldest and largest global environmental network, IUCN is a democratic membership union with more than one thousand government and NGO member organizations, and some ten thousand volunteer scientists in more than 160 countries. http://cms.iucn.org.

Mad City Chickens. Community ordinances regulate chickens within city limits. This site for chickens in Madison, Wisconsin, has links to legal language and pertinent arguments. www.madcitychickens.com.

Information to answer the questions that arise when raising chicks like these Dominique youngsters is available in books and on the Internet. Consult local agriculture organizations such as the FFA Organization and county extension agents for advice. *Bryan K. Oliver*

POULTRY BREEDS

Bird Studies of Canada. Studies migratory swans. www.bsc-eoc.org.

Cook's Peacock Emporium. The owners of this Alabama peacock farm are active in the United Peafowl Association. www.peacockemporium.net.

Environmental Studies on the Piedmont. Conducts research on wild swans. www.trumpeterswans.org.

Game Bird Gazette. www.gamebird.com.

Jeannine Ferguson. Guinea Fowl Breeders Association and other supportive material. www.guineafowl.com.

Holderread Waterfowl Farm & Preservation Center. Offers rare domestic ducks and geese in more than sixty varieties. www.holderreadfarm.com.

Lou Horton. This distinguished waterfowl breeder provides information from his long experience. www.acornhollowbantams.com.

Metzer Farms. This California duck and goose hatchery provides information on ducks, geese, guineas, game birds, and related products. www.metzerfarms.com.

Mouse Creek Feather Farm. The owners of this Wisconsin peacock farm are active in the United Peafowl Association. www.mousecreekfeatherfarm.com.

National FFA Organization. Dedicated to agricultural science education and making a positive difference in the lives of young people. www.ffa.org.

National Wild Turkey Federation. A nonprofit organization dedicated to conserving wild turkeys and preserving hunting traditions. www.nwtf.org.

North American Gamebird Association. A nonprofit organization dedicated to improving methods of game bird production and hunting preserve management. www.naga.org.

Pheasants Forever. A hunters' advocacy organization for pheasants and quail, under its sister organization, Quail Forever. Its site lists state wildlife agencies governing game birds. www.pheasantsforever.org.

Shady Hollow Farm. This Maine farm raises large fowl and bantam chickens, peafowl, guinea fowl, turkeys, quail, partridges, and pheasants. www.shadyhollowfarm.com.

Society for Preservation of Poultry Antiquities. Provides information on all breeds, recognized or not. www.feathersite.com.

University of Michigan Museum of Zoology. Animal Diversity Website gives information on pigeons and doves. http://animaldiversity.ummz.umich.edu.

USDA National Agricultural Library. www.nalusda.gov.

PREDATORS

Animal tracks. Identify your predator. www.mass.gov/dfwele/dfw/dfwpdf/dfwtrax.pdf.

Humane Society of the United States. Advice on non-lethal means of dealing with predators. www.hsus.org/wildlife.

Internet Center for Wildlife Damage Management, a collaboration of Cornell University, Clemson University, University of Nebraska–Lincoln, and Utah State University. http://icwdm.org/Inspection/livestock.asp.

Livestock guardian dogs. http://www.lgd.org.

The National Sustainable Agriculture Service. Advice on predators. http://attra.ncat.org/attra-pub/predator.html.

PRODUCTS

Down, Inc. The feather and down business of Eurasia Feather Company. www.downinc.com.

Food Timeline. Background information on food history and cooking. www.foodtimeline.org.

Heritage Foods USA. Food broker connecting producers with purchasers. www.heritagefoodsusa.com.

LB Processors. A state-approved facility producing emu oil. www.lbemuoil.com.

Maple Leaf Farms. The major commercial duck producer. www.mapleleaffarms.com.

Ostrich Growers Meat Company. California ostrich ranch. www.ostrichgrowers.com.

Slow Food USA. Nonprofit organization championing food that is good, clean, and fair. www.slowfoodusa.org.

Turkeys are welcome in the tobacco fields at Claude Moore Colonial Farm in Virginia. These Black Spanish turkeys would have been a familiar sight during the eighteenth century. Domestic turkeys could be driven to market as a flock, feeding on acorns and insects along the way. *Jan Tilley, Claude Moore Colonial Farm*

TOXIC PLANTS

Cornell University. The Poisonous Plants Informational Database lists plants and their effects on livestock and other relevant sites. www.ansci.cornell.edu/plants.

Red Oak Farm. Allen and Myra Jones Charleston have a page on dangerous plants on their site with much helpful information about emus. www.redoakfarm.com.

VETERINARY

Association of Avian Veterinarians. Provides a veterinarian-finding service. www.aav.org.

Lafeber Company. Premium pet bird food and toy store that offers a veterinarian identification service. www.lafeber.com.

National Association of State Departments of Agriculture. Assists in finding state department of agriculture phone numbers and addresses in a state-by-state directory. www.nasda.org.

United States Animal Health Association. Lists state veterinarians. www.usaha.org.

Courtesy of Corallina Breuer

Christine Heinrichs has been writing about poultry since the 1980s. She and her daughter bought some chicks at the feed store one day, the first step on their poultry path. The Society for the Preservation of Poultry Antiquities helped her learn about the beautiful chickens that amazed and delighted them. As a professional writer, she became volunteer publicity director in 2000. Heinrichs maintains a website devoted to poultry: www.poultrybookstore.com.

Her degree in journalism from the University of Oregon makes her a Duck. She is a member of the Society of Environmental Journalists and the Society of Professional Journalists.

She is the author of *How to Raise Chickens* and has written about business, engineering, golf courses, and law schools, as well as poultry and other animals. She is a docent for the Friends of the Elephant Seals, greeting visitors to the Piedras Blancas rookery and explaining about these unusual marine mammals when she isn't watching chickens.

She lives in Cambria, California, with her husband Gordon, who knew what he was in for when they spent their first date watching chicks hatch. She looks forward to Dorkings in her future.